DARWIN

Also by Niles Eldredge

Why We Do It: Rethinking Sex and the Selfish Gene

Reinventing Darwin: The Great Debate at the High Table of Evolutionary Theory

Life on Earth: An Encyclopedia of Biodiversity, Ecology, and Evolution

The Triumph of Evolution and the Failure of Creationism

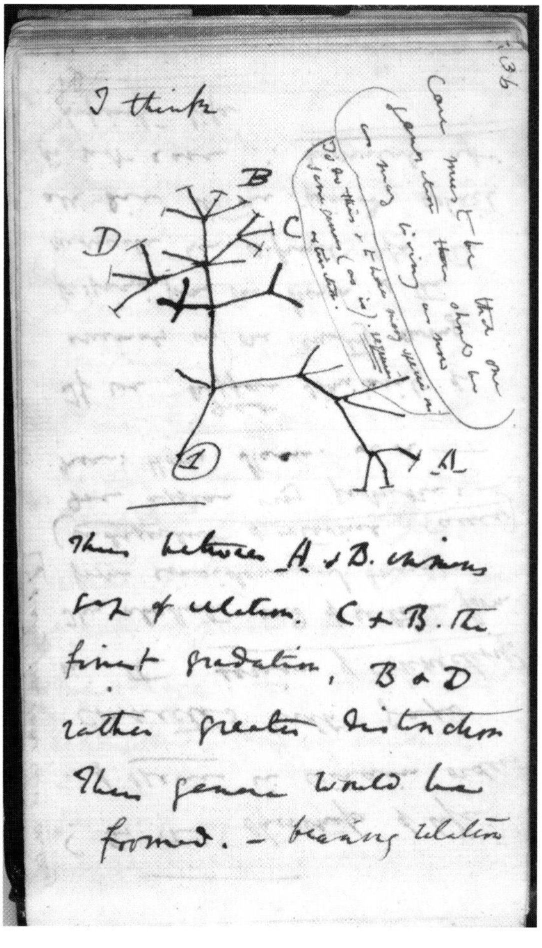

36

I think

Case must be that one generation then should be as many living as now. To do this & to have many species in same genus (as is) requires extinction.

B

C

D

1

A

Thus between A & B immens gap of relation. C & B the finest gradation, B & D rather greater distinction Thus genera would be formed. — bearing relation

Niles Eldredge

W. W. NORTON & COMPANY
New York • London

Printed in the United States of America
First Edition

For information about permission to reproduce selections from this book, write to Permissions, W. W. Norton & Company, Inc., 500 Fifth Avenue, New York, NY 10110

Manufacturing by RR Donnelley / Crawfordsville, IN
Book design by Rhea Braunstein
Production manager: Julia Druskin

Library of Congress Cataloging-in-Publication Data

Eldredge, Niles.
Darwin : discovering the tree of life / by Niles Eldredge.—1st ed.
p. cm.
Includes bibliographical references.
ISBN 0-393-05966-9 (hardcover)
1. Darwin, Charles, 1809–1882. 2. Naturalists—England—Biography.
3. Evolution (Biology) 4. Natural selection. I. Title.
QH31.D2E43 2006
576.8'2'092—dc22

2005018636

W. W. Norton & Company, Inc., 500 Fifth Avenue, New York, N.Y. 10110
www.wwnorton.com

W. W. Norton & Company Ltd., Castle House, 75/76 Wells Street, London W1T 3QT

1 2 3 4 5 6 7 8 9 0

To Stephen Jay Gould (1941–2002),
who also loved Charles Darwin

CONTENTS

LIST OF ILLUSTRATIONS

DARWIN

Introduction

The 200th anniversary of Charles Darwin's birth and the 150th anniversary of the publication of his epochal book *On the Origin of Species* both fall in 2009. Many events are planned on both sides of the Atlantic to celebrate this creative genius who founded modern evolutionary biology.

The first of these celebrations is a paired exhibition and book on Darwin's life and work, including an up-to-date display on modern evolutionary biology. I am the curator of the exhibition, responsible for its scientific content; this book reflects my thinking about Darwin's life and work, in parallel with the contents of the exhibition.

Books and exhibitions are vastly different media. Exhibitions justly rely on three-dimensional objects—specimens, artifacts—to convey the message, with a minimum of accompanying written explanation. Books are the opposite: though I have endeavored to capture the exhibition's contents with the illustrations in this book, its real strength lies in the opportunity to set forth a more complete analysis and explication of the central ideas on Darwin that are common to both projects. Then again, the Darwin exhibition is an institutional product—while I alone am responsible for the contents of this book.

These central ideas concern Darwin's thought processes—his creativity—and include the evidence that led him as a young creationist still in his twenties to become convinced that all of life has descended from a single common ancestor in the remote geological past: in short, evolution. The young Darwin was highly intuitive—letting nature come to him in almost impressionistic ways. He himself said that he was a true Baconian inductivist, and the present narrative explains that he was right in thinking of himself in this way, contrary to the claims of many scientists and historians who have concluded that Darwin was really *not* the inductivist he said he was.

But it is also true (as again Darwin himself said it was) that he was highly analytic, and an early exponent of what has come to be known as the "hypothetico-deductive" approach to doing science. The main conclusion of my several years of research into Darwin's life and work for both the exhibition and this book is that when Darwin reached home after five years on the *Beagle*, he had become convinced of evolution through his experience of several different patterns he encountered in the fossils and living species of South America, and of course through his observations of species on the various islands of the Galapagos Archipelago. He then reformulated these patterns as predictions—and tackled the literature and began quizzing correspondents from around the globe to test his predictions, as well as three additional predictions he formulated while living in London in the late 1830s. The final step, after his discovery of natural selection, was to rethink those patterns/clues/predictions, deriving them afresh from his concept of how natural selection works.

All this was achieved in a few short years—largely between 1837 and 1842 (though the *Origin* was not to be published for another seventeen years). The "evolution of Darwin's evolution" is remarkably preserved in a series of notebooks and manuscripts not published in Darwin's lifetime. These documents allow an extraordinarily close look at the essence of creativity. They show how similar creativity in science truly is to all other facets of the human creative experience. They also reveal, when compared to his views as finally published in the *Origin*, how Darwin abandoned some of the early patterns that led him to the conclusion that life has evolved in the first place—abandoned simply because they seemed to him inconsistent with his understanding of the principle of natural selection. Darwin gave us the core of evolutionary theory today. But in downplaying the significance of some of his crucial early observations (for example, on the importance of isolation and on the mode of replacement of species in the fossil record), he put something of a stranglehold on evolutionary biology, which has lasted in some circles to the present time.

My analysis of Darwin's notebooks and unpublished manuscripts forms the heart of this book—and adds insight, I believe, into achievements that have not been glimpsed, at least in such detail, before. Though the accompanying exhibition includes these "crown jewels," there is no way that their significance can be as deeply appreciated in the context of an exhibition as they can in a book. That is why books and exhibitions are both necessary, and in this case truly complementary to each other.

Both the exhibition and the book take on added significance simply because evolution, once again, is seen as a controversial idea. We increasingly hear that evolution is "only a theory." Statements to that effect are even being added as warning labels in high school textbooks. It is important to combat this rising tide of what seems like willful ignorance. We need to teach our kids what science is and how it works. We need to spread the word that *all* the great conclusions of science are theories: quantum mechanics, gravitation, plate tectonics, special relativity, the nature of light, etc., etc.; all these are theories. Theories are complex sets of ideas about the nature and workings of natural phenomena. They are tested over and over again—and take their place as accepted virtual fact in science, with the proviso that an alternative formulation may eventually emerge that fits the observations just a little bit better.

Hence evolution: the evidence that life has evolved is so overwhelming that no serious biologist since Darwin has disputed it. Evolution is as firmly established as is the notion that the Earth is an oblate spheroid that spins on its axis and rotates around the Sun. Yet that picture of the relationship among components of the solar system is also a theory—though no one would say it is "only a theory."

Thus the ultimate aim of this book and the exhibition with which it is paired is the clear demonstration of the evidence and mode of thinking that led Charles Darwin to the conclusion that life has evolved.

Acknowledgments: The exhibition assembles a truly unprecedented array of Darwiniana from the major repositories of manuscripts, specimens, and personal effects in England, some of which have not been together since Darwin was on board HMS *Beagle* in the 1830s. The exhibition opens in New York at the American Museum of Natural History in November 2005; it then travels to the Boston Museum of Science, Chicago's Field Museum, and the Royal Ontario Museum in Toronto, before opening at London's Natural History Museum in time for Darwin's birthday celebration on February 12, 2009. I am very proud that the initial idea for the exhibition came from the institution where I have spent my entire professional career: the American Museum of Natural History. I am grateful to the museum's president, Ellen V. Futter, and to its provost, Michael Novacek, for giving me the responsibility for developing the exhibition's content.

No American institution could, of course, possibly mount a major exhibition on Charles Darwin without the full, enthusiastic support of key persons and institutions in Great Britain. I am deeply grateful especially to Randal Keynes, a great-great-grandson of Charles Robert Darwin, co-founder of the Darwin Trust, and himself a Darwin scholar and author. Randal has been of vital importance in connecting our exhibition team up with the major public and private holdings of Darwin manuscripts, specimens, and artifacts. He has cooperated with enthusiasm and great imagination in the development of the exhibition. On a personal note, Randal brought me up to speed very quickly on many points about Darwin's personal life, as well as his scientific work, and he has read this manuscript for me. In a profound sense, Randal Keynes gave me the encouragement, hence confidence, that I needed to take on the daunting task that is Charles Robert Darwin.

Through Randal, the exhibition team and I have been introduced to other members of the Keynes family, all of whom have been most generous with their time, advice, and permission to utilize Darwinian objects in the family possession: to Richard, Stephen, and Simon Keynes, my heartfelt gratitude.

Thanks, too, to the staff of English Heritage, especially to Julius Bryant and to Tori Reeve, the Curator of Down House in Kent, where Darwin spent the last forty years of his life. And also to the staff of the Departments of Palaeontology, Zoology, and Botany at the Natural History Museum in London—and especially to Gordon Rankmore, Christopher Mills, and Sandra Knapp of that excellent institution.

Cambridge University, which is the repository of so much of Darwin's correspondence and early unpublished manuscripts, but also the place where so many of Darwin's zoological, botanical, and mineral specimens from his epic voyage are housed, has been crucial to the success of our venture. Thanks especially to Adam Perkins and Patrick Zutshi (Cambridge University Library), Adrian Friday and Michael Akam (Zoology), and John Parker and Gina Murrell (Botany).

A wise person approaches the world of Darwinian scholarship with some trepidation. After all, the "Darwin industry" has long been established, peopled by superb scholars. In addition to Randal Keynes, I am very grateful to two historians, Pietro Corsi and James Moore, for enlightening discussions. And I am particularly grateful to David Kohn, botanist and Darwin historian, for his direct and extremely valuable contributions to the production of the exhibition—and perhaps even more for his candid discussions about Darwin's thinking in those creative formative days on the *Beagle* and above all upon his return to London. It was David Kohn who transliterated and annotated Darwin's "Transmutation Notebooks," literally the work that made the present study possible.

My colleagues and friends William Miller III, Harold B. Rollins, and John Thompson did me the honor of reading the manuscript, and provided many valuable suggestions.

The Department of Exhibition at the museum is superb, under the leadership of David Harvey. The Darwin team has included Geralyn Abinader, Harry Borelli, Dempsey Collins, Sarah Galloway, Ellen Giusti, JoAnn Gutin, Lauri Halderman, Tim Nissen, Melissa Posen, Steve Quinn, Stephanie Reyer, Martin Schwabacher, Joel Sweimler, Robert Vinci, Mindy Weisberger, Julia Wilk, and Sarah Wilson. I am truly grateful to them all for a terrific experience in mounting the exhibition. Thanks also for the input of the staff of the collaborating institutions to which the exhibition travels after its opening in New York.

I would like to thank my agent, John Michel, my editor, Angela von der Lippe, and her assistants, Alessandra Bastagli and Vanessa Levine-Smith, for all their help in making this book take shape. I am grateful too to Vikii Wong, who tackled the arduous task of photo researching; it has proved especially difficult to match book illustrations with exhibition contents given the production schedules of these two very different, albeit complementary projects.

My friends and colleagues at the American Museum, Sid Horenstein and Ian Tattersall, have provided their usual good cheer and timely advice throughout the entire gestation period of book and exhibition. May we continue in like manner for many years.

And finally, my best friend and life partner, my wife Michelle J. Eldredge, who took on the role of researcher when I felt most at sea confronting Darwin's life and work. We sat in cafés in New York, London, and Paris and contemplated Darwin's life, work, and meaning for modern times. She read extensively and found out the answers to many nagging questions. Her work especially on a crucial field trip (with David Harvey and Randal Keynes) to Down House, Cambridge, and the Natural History Museum in London in early 2004 proved invaluable. I love her, thank her—and hope there will be many more exciting ventures to come.

CHAPTER 1

Charles Darwin

Charles Robert Darwin was born on February 12, 1809—the very same day as Abraham Lincoln. Both men shook the world in their own time. Both detested slavery. And both now have their images on the low-denomination (hence more common) paper currency of their natal country. Lincoln, of course, is on the five-dollar bill. Darwin is on the British ten-pound note—where he replaced another prominent Victorian figure, Charles Dickens, in 2000.

The ranks of towering nineteenth-century figures who retain a firm grasp on twenty-first-century collective consciousness are rapidly thinning. Darwin replaced Dickens on the ten-pound note, it was joked, because he had the nicer beard. But the truth is that Darwin, like Lincoln, retains a relevance to life in the twenty-first century that Dickens has all but lost. So too have Karl Marx and Sigmund Freud begun to fade—astonishing given the amount of controversy they both engendered well into the latter half of the twentieth century. But with the collapse of the Iron Curtain and the advent of Prozac, society at large seems to have metabolized these two giants to the point where mere mention of their names fails to stir much passion one way or the other. Indeed, their names are simply not called as much as they once were.

Lincoln survives as the savior of the Union—and symbolizes at home and throughout the world the deeply humanitarian impulse that is still far from fully realized in the modern body politic. He is an icon of hope for social decency as yet not realized, yet thankfully far from forgotten.

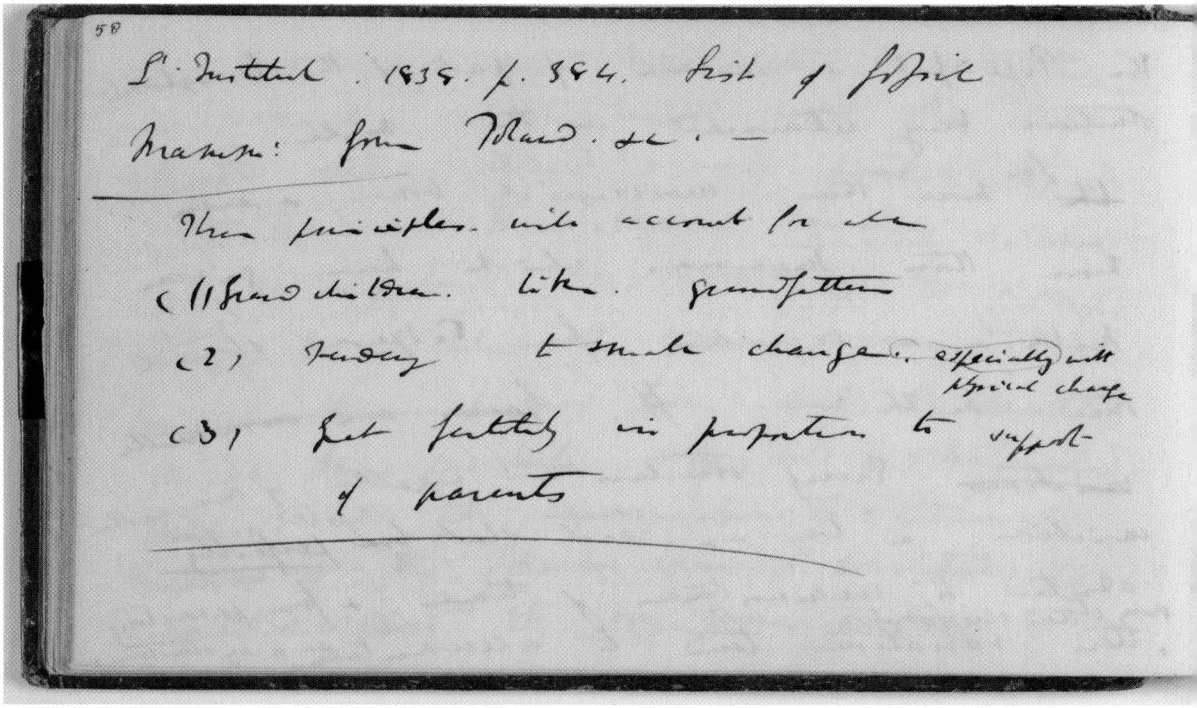

So, too, with Charles Darwin. In a recent poll in Great Britain, in which BBC viewers voted on their top ten all-time British figures, Darwin came in fourth, behind Princess Diana (third place; Winston Churchill was first, and a Bristol train bridge engineer, Isambard Kingdom Brunel, was second). The reason why Darwin placed higher than William Shakespeare, Isaac Newton, Elizabeth I, and Queen Victoria (the possibilities go on and on) must surely be that Darwin, like Lincoln, has not been fully absorbed into the mainstream fabric of our collective Western culture. He remains, like Lincoln, a symbol of a worldview—a view of what life is and how it came to be, and above all else, a view of who we humans are and how we came to be—that remains in some quarters a promise yet to be fully realized, and to others a satanic threat against all that is good and holy.

For what Darwin did was to transform the prevailing view of stability—of the earth, of all the species on earth, and not least the stability of society's strata—into a picture of motion. Let enough time go by, and the laws of biological transformation will inexorably and inevitably transform life on earth. Old species will go, replaced by

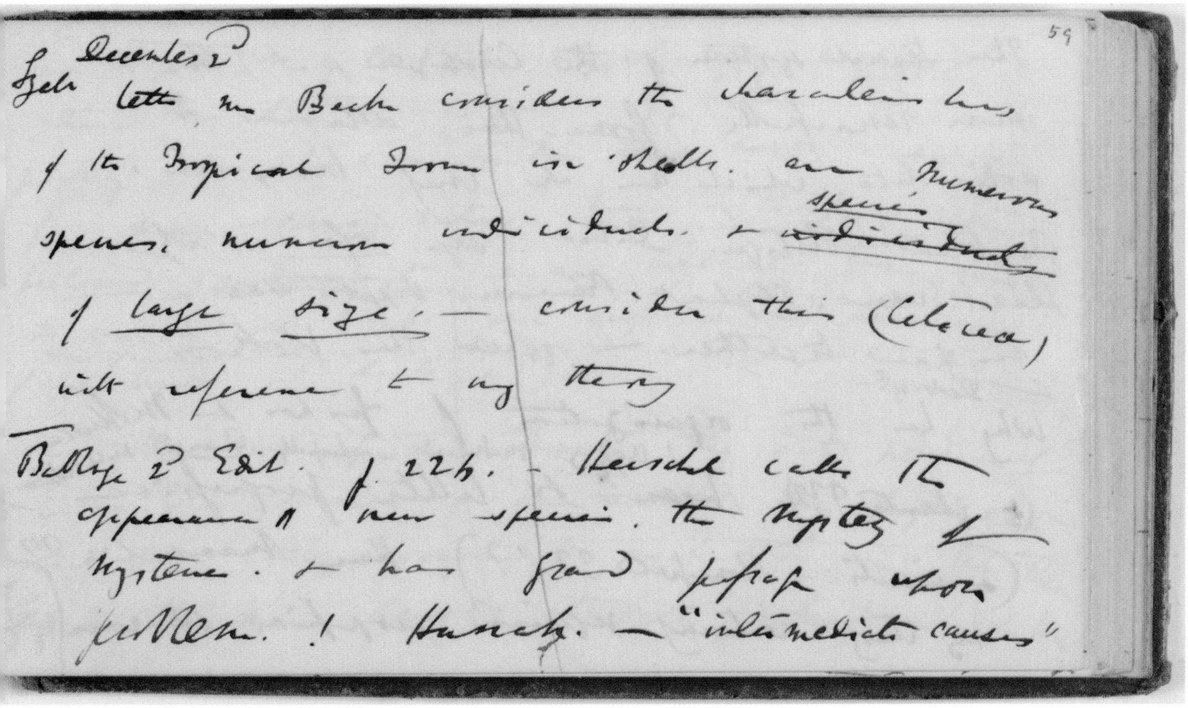

Passage from Darwin's Notebook E, of 1838–39, where Darwin exults over the publication of John Herschel's letter to Sir Charles Lyell. In that letter, Herschel wondered when a naturalist would appear to solve the "mystery of mysteries": how similar species come to replace extinct ones in the fossil record. "Herschel calls the appearance of new species, the mystery of mysteries. & has grand passage upon problem! Hurrah." The left-hand page contains a pithy summary of his theory of natural selection: heritable variation and geometric increase of populations.

new ones. John Herschel, perhaps England's leading scientific light in the 1830s, openly wondered when a naturalist would emerge who would develop a plausible scientific answer to what Herschel called the "mystery of mysteries"—how it was that extinct species are replaced by new ones as the geological ages roll by. Tradition had it that new species could only be created by God.

Darwin was that naturalist—that "Newton of a blade of grass," as Immanuel Kant put it. As we'll see in more detail, some of his predecessors had already suggested that the earth is vastly older than the ten thousand or so years that biblical accounts seem to imply, and that, given enough time, the sorts of processes, like erosion and deposition of sediments, or earthquakes and lava flows, happening now on its surface, could transform the earth as deeply and radically as Darwin later showed could happen for organisms. The tools necessary to read the pages of earth's history from the rock outcroppings dotting the landscape were already emerging.

And Darwin had his evolutionary predecessors, too, including his own grandfather Erasmus Darwin, daring to suggest a process of descent linking up all forms of life. Even as they remained for the most part creationists, Darwin's immediate forerunners were beginning to map out the patterns of what they ambiguously called "affinity" among the myriad species of life on earth.

So Darwin didn't invent the idea of evolution. He himself seems to have been an ordinary creationist when he set forth on the "most important event of my life"—the five-year *Beagle* voyage around the world. As we shall see, by the time he got back, he *was* an evolutionist. That was in 1836.

Darwin spent the rest of the decade thinking evolution through—coming up, by 1838, with the central mechanism of evolution: natural selection. Yet it took him until 1844 to mention his views, very tentatively, to anyone outside the family. And he never published a word (though he did drop a few hints now and again) until pushed to do so by the fateful arrival, in June 1858, of a letter and manuscript from Alfred Russel Wallace, outlining something so close to what Darwin had long since come to think of as "my theory" that his heart sank.

Why? Why had Darwin waited twenty years—despite his ambition to "take his place among men of science" (which he meanwhile managed to do in other ways, especially through his geological expertise)? For he knew his idea was truly earth-shattering and would, as he foresaw, transform the way science looks at life on earth.

It was that stabilist view, the very one he grew up in, that made him hesitate. The men of science in the early nineteenth century adhered to it, and Darwin knew that to do the job right, he had to have ironclad evidence and a convincing explanation of how evolution happens. But that was not all; Darwin must have felt that he had the Victorian equivalent of the plans for an atomic bomb in his brain, so certain was he of the social opposition he knew "my theory" would face.

But he was an independent man of property, a man of inherited means. Why should he care what the laypeople of Britain might think? The real key to Darwin's dilemma lies in the fact that most of British society adhered to the biblical view of the origin of the world and of life—and the stability of all things more or less since their creation—as outlined and upheld especially by the Church of England. This encompassed people from all walks of life—including, critically, the academic world. Everyone—or nearly everyone—was a creationist. There was little in the way of a professional class of scientists when Darwin came along, though by midcentury, such a class of professionals had begun to emerge. Darwin was eventually to find much support among them, so that for those entering the profession after 1859, the revolution in thinking had been completed in science.

But Darwin's mentors and older colleagues from the 1820s into the 1850s were either the monied elite or clergymen. All religious, all vested in the status quo. And the status quo meant first and foremost social stability. The upper classes were favored by a vestigial form of "divine right"—they deserved to be the elite simply because that was the way things were, the way God wanted it. How ironic that a parody of Darwin's own ideas of natural selection was later to become the new way of justifying the existence of the elite: the thought that cream rises to the top, and that the people with all the money and power occupying the highest social strata are there purely because they deserve to be—they earned it, surviving in the dog-eat-dog competitive world of "social Darwinism."

So Darwin was not just against a prevailing scientific view—for the scientific view, echoing the commonsensical perception that the world and its species is indeed stable, fit in so nicely with prevailing religious views that to attack the one was to attack the other. Indeed, the "scientific" view was essentially a religious view: the rising tide of rationalism—in science, such achievements as Newtonian physics and the beginnings of chemistry—had yet to influence ideas on the living world. Religious doctrine dominated biological thinking.

And that, as we'll see as we look at Darwin's earliest private notebook jottings on evolution, was precisely how he saw the problem. The only alternative game in town—as Darwin saw all too clearly—was biblical, creation-imbued "science." Its tenets included the belief that the earth was only ten thousand years old (though some of his predecessors, including his Cambridge mentor, the clergyman Adam Sedgwick, were already shooting holes in that argument). Species might become extinct through natural causes, but new species take their place by individual acts of creation, divine intervention. That was the main line of thinking in Darwin's youth—a "scientific" view so closely tied to received religious wisdom that it was impossible to tell where the one started and the other left off. And (Darwin's deepest secret) he immediately extended his evolutionary ideas to the origin of our own species, *Homo sapiens*.

Darwin split the scientific world off from the religious world, at least as far as the "mystery of mysteries" was concerned. He convinced the scientific world—and beyond that, the thinking world—that life has evolved through natural causes. Because he was so careful in building the case for evolution, and for the evolution of human beings, Darwin did more to secularize the Western world than any other single thinker in history. He was right to anticipate the howls of outrage—and would not be surprised (though perhaps disappointed) to learn that the outrage continues unabated in many quarters of the Judeo-Christian world.

And that is why Darwin's face is on the ten-pound note. Even though creationism is a less vociferous strain in modern British life than it is in many of the former colonies (the United States, for example), in his native land there are many people still very uncomfortable with the apparently anti-religious nature of the idea of evolution. The fact that evolution is an idea that challenges ancient Middle Eastern ideas about the history of the earth and its species rather than the core of religious belief doesn't matter. Rather, the idea of evolution challenges received wisdom on how we humans got here, and thus who and what we are, and *that's* what matters.

Darwin permeates society. I recently found his name invoked twice in the same issue of *The New York Times*, where it was used as shorthand for down-and-dirty competition in the business world and, elsewhere, as a synonym for evolution in an account of creationist struggles in the American Midwest. "Darwin" or "Darwinism" simply means "evolution"—perhaps not so much in biology these days, but certainly in American society at large. The famous "Battle of the Fishes" played out on bumper stickers and refrigerator magnets all over the United States is a case in point. The Christian symbol of the fish, of course, began it all. Then someone reinforced the point by writing "Jesus" inside the fish's profile. Then a *provocateur* replaced "Jesus" with "Darwin"—starkly pointing to the social dichotomy that persists to this day. The next step: someone added feet to the fish, recalling the myriad cartoons of fish-to-man that have become the hallmark of popular Darwinian iconography.

But my favorite bumper sticker in these Darwin wars eschewed the fish entirely, depicting instead a graveyard with a tombstone marked "R.I.P. Charles Darwin" and proclaiming that now Darwin finally knows the truth. For years the rumor persisted that Darwin recanted evolution on his deathbed (he died of a heart ailment in 1882 at the age of seventy-three). There is absolutely no documentation of such a renunciation and every reason to agree with historians who have concluded that Darwin had become an agnostic (if not an atheist) decades before he died. For years he had professed a residual belief in the Almighty; but the death of his ten-year-old daughter

Annie in 1851 is widely believed to have removed whatever lingering religious sentiments he may have had by then. Annie's death also seems to have helped steel his resolve at long last to publish his "theory." Though his public position was always more or less along the lines of "agnosticism" (a term coined by his so-called bulldog, T. H. Huxley), long before his death Darwin had abandoned active religious faith entirely.

And so the die was cast—and the echoes of this pitched battle resonate today. Poll after poll shows the American public as evenly split over "Darwinism" as it is over the Republican "compassionate conservatives" versus the Democratic "liberals." There is even a strong correlation between the two, as the Christian right of course constitutes the core support of the Republican Party these days. But it is also the case that 40 percent of scientists still proclaim themselves religious—the same percentage as found in a poll way back in 1916. And many mainstream Protestants, Catholics, and Jews have had no trouble reconciling their faith with modern science, including genetics and evolutionary biology. As Darwin himself saw, there is no *necessary* conflict between the two. Only the insistence on strict biblical literalism—the belief that everything in the Bible must be true, including the mutually inconsistent accounts in Genesis on the origin of the earth, life, and human beings—forces a collision between the worlds of biology and Judeo-Christian religion.

So, creationism persists. The sole difference between the current version and that of the first half of the nineteenth century is that today's creationism has been expunged from the scientific account. We'll return to these themes in the last chapter.

Darwin—poor old Darwin—is also invoked in other arenas of social thought and even policy. Just as Marx is said to have exclaimed, "*Moi, je ne suis pas Marxiste* (I am no Marxist)," Darwin would cringe at some of the movements undertaken in his name. The same so-called social Darwinism that gives us an approving image of a dog-eat-dog competitive world has also given us the eugenics movement and some of its darker outgrowths, such as the genocidal practices of the Nazis in World War II—where eugenics was invoked as a scientific rationale to go along with whatever other "reasons" Hitler and his fellow Nazis had for the Holocaust. The eugenics movement can actually be traced in part to Francis Galton, a cousin of Darwin's. The central idea is to improve the human race by not allowing people with hereditary defects to breed. The ethical problems are insurmountable: for one thing, what is a "defect"? Are my missing premolars and bad eyesight enough to ban me from breeding (too late, in any case!). And who gets to decide?

Darwin saw a "grandeur in this view of life," meaning that, in evolution, he saw a simple natural process that underlay the production of all the myriad, fabulous, even downright beautiful animals and plants in the world. Yes, he saw nature as competitive—and often cruel. But he had an equally well-developed aesthetic appreciation of the natural world. Small wonder, then, that ethicists of many stripes—philosophers and evolutionary biologists, even natural history–loving theologians—have tried to use evolution as a springboard for deriving a system of ethics in keeping with their understanding of the nature of the evolutionary process.

Unsurprisingly, there turns out to be no neat, one-on-one correspondence between evolution and any single system of ethics that can be derived from it. The great Russian geneticist Theodosius Dobzhansky saw evolution as wholly compatible with the central tenets of Christian belief. Others see nature red in tooth and claw—and write that we must fight against our baser, evolutionarily based, gene-driven propensity to compete with our fellows. The ethical homilies you draw from evolution depend upon which aspect of the evolutionary process you focus on. Darwin, I think, would be bewildered by the chaos of conflicting ethical "conclusions" that have been drawn in his name.

Yet nearly everyone seems to think that it does make a moral, ethical difference whether we were specially created by a benevolent God or whether we evolved from ancestral species along with the rest of all living things. But does it? It makes a huge difference in the actual details of the story we tell ourselves of how we came to be—and in that sense who we are. But as human beings we make our own laws to govern our conduct, laws based in part on ethics. These can be seen as pragmatic rules of behavior to stabilize social life—a view I personally think is more defensible than deriving codes of behavior based on the supposition either that we were created in God's image or that we evolved through natural processes. And though I cannot be sure that Darwin would agree, I do think he would be relieved not to be seen as the destroyer of traditional ethics.

One impact that Darwin has undeniably had on society—or at least on those of us who ponder his ideas and all their ramifications—is to re-forge the emotional and conceptual bonds of humanity with the rest of life. He serves as an anchor to the natural world for a humanity that, ever since we invented agriculture, has become increasingly divorced from the rest of the living realm. No longer reliant on the natural fruits of nature for our sustenance, over the past ten thousand years, naturally enough, we have come to see ourselves as living outside of—and even superior to—the rest of nature.

And this change in our ecological status, I believe, is reflected in the "dominion" passage in Genesis: the earliest farmers, including those who drafted those marvelous ancient texts that survive as the Pentateuch, or first chapters of the Old Testament, knew full well they were animals, but animals with a difference. They saw they were not in fact a part of nature in the same way as the birds, mammals, reptiles, and fishes around them. Their explanation of how all this came to be—that God created everything, but created humans in His image—takes this ecological difference between humans and all other forms of life into account very well.

Nor does it seem far-fetched to see that humans as hunter-gatherers (as we know from anthropological accounts of surviving hunter-gathering groups) saw their gods as manifestations of part of the local ecosystems in which they saw themselves as participants. Humans left their local ecosystems when they invented agriculture, and needed *something* to feel part of—such as a cosmos directed by a single, Almighty God. According to this line of thought, it was we who created God in our image.

Whatever the truth of all that, it is very much the case that agriculture has triggered the expansion of the global human population from an estimated 5 or 6 million a scant ten thousand years ago to well over 6 billion early in the twenty-first century. Increasingly, we live in cities, getting our food from stores, our water from the tap, our energy sources from flipping a switch. Our explosion in numbers all over the globe has begun to drive many species extinct (Darwin would be heartbroken); and though there are plenty of pragmatic reasons to try to curtail this rampant extinction, Darwin's lesson that we came from nature and so in many ways are still a part of it (even though we have changed the rules of engagement with it) is indeed an anchor that stabilizes our vision of who we are and what we are doing to the earth and all its inhabitants.

My favorite piece of social Darwiniana makes this very point in ways far better than I can do. The Mexican painter Diego Rivera was once commissioned to produce a mural in the foyer of Rockefeller Center in New York. He created *Mankind at the Crossroads*—a typically busy Diego piece, crowded with over 175 people, many of them recognizable historical figures. At the center of the piece sits a man at the controls beaming out with rays filled with images of the macrocosmos and the microscopic world—a world of modern scientific knowledge. Around him are warring parties: Communists, Nazis; warring philosophies and religions. His original version had an image of Lenin—to the dismay of the Rockefellers, who ordered him to remove this symbol of communism from their foyer. Rivera refused, so they paid him off and destroyed the mural.

Diego re-created the work the following year (1934) in the Palace of Bellas Artes in Mexico City. He threw in an image of a martini-drinking, angry-looking John D. Rockefeller, Jr., for revenge—and added some more Communists for good measure.

But as arresting as this entire mural is, it is the image of an infinitely wise and perhaps rather sad-looking Charles Darwin at the lower left corner of this huge painting that is the most riveting. For Darwin is absolutely the *only* person in the painting to look the viewer straight in the eye. You can walk the entire length of the mural, and examine everyone else from any and all angles, and they all manage to avoid catching your eye. Except Darwin.

Darwin is looking at you, and with his left hand he points at a monkey, who is holding the hand of a human child. There are other sorts of animals nearby and also a tankful of sea creatures. A little further toward the center of the mural lies a major tableau of Diego's beloved native agricultural products of Mexico.

It is as if Rivera has Darwin telling us that, whatever you might make of the implications of science and technology in the modern world—and whatever you might think of how science fits into the existing warfare between philosophies, social movements, and religions—one thing is true: We are of the earth, and we are connected to absolutely all other living things on the planet. That, at least, is one thing we know as the truth, and should never lose sight of.

But Darwin, amazingly, is at the core of the modern experience scientifically as well. Theodosius Dobzhansky once wrote that "Nothing in biology makes sense except in the light of evolution," a statement that remains as true today as it ever was. Darwin looked in vain for an evolutionary mechanism in the principles of heredity as they were then (largely incorrectly) understood in the 1830s and 1840s; but when he formulated the principle of natural selection, he had discovered the central process of evolution. Nothing that we have learned in the intervening 175 years—about the structure and function of DNA and RNA, all the sophisticated modern understanding of genetics—has contravened Darwin's basic description of how natural selection works. Nor have we changed our views on how artificial selection serves as both a guide for understanding selection in nature and a technique for the human manipulation of genetically based features of domesticated and laboratory organisms.

People have been manipulating the genes of organisms for at least 15,000 years, probably longer. Until recently they had no knowledge of why it works; but if you just let those organisms—plants or animals—breed, the ones that display the qualities you value the most (more milk in cows, more gentle behavior in dogs, etc.), you are favoring the passage of the genes that yield those results to the next generation. All of our agricultural products—plants as well as animals—have been developed through this process. Now that we have genetic engineering, we can short-circuit the process, cutting more directly to the chase of determining what genes and gene combinations will be selected. But it is still selection.

We also see selection in action in the constant battle to develop drugs resistant to diseases; the disease pathogens (for example, the AIDS virus, or the malarial *Plasmodium* parasitic protozoan) are constantly mutating—with strains resistant to whatever form of drug is in use being instantly selected as the surviving form of the disease. This is *very* Darwinian, this menacing drug-resistance dance we humans have with our killer diseases.

But why should Darwin's name still be invoked even in science? After all, science marches on—and as we shall see, we have learned a great deal about evolution that Darwin knew nothing about. On the other hand, Darwin staked out the evolutionary territory so broadly and so thoroughly that, in a general sense, he did literally define the entire content of evolutionary biology right down to the present day. It is as if he saw the relevance of whole areas of inquiry, with no way of knowing the details that have emerged. He sought mechanisms of evolution in the principles of heredity, with absolutely no knowledge of genes and their behavior; but in so doing, he managed to demonstrate the relevance of heritable variation once it became more thoroughly understood. He saw the process of development of adults from fertilized eggs, itself a process of transformation, as crucial to understanding how anatomy changes in evolutionary time—again, without a glimmer of understanding of the molecular genetic basis of the developmental process. He saw the importance of the environment in the process of natural selection and adaptation. He realized the importance of geography in isolating parts of species that were to go on and become new species. And of course he discovered natural selection (as did Alfred Russel Wallace independently), still seen as *the* quintessential evolutionary process.

But there is more: Darwin is as fresh today as he ever was because we evolutionary biologists have yet fully to realize the value of everything he accomplished. I was astonished to find how modern some of Darwin's early thoughts were: as a single example, his principle of divergence is hardly noticed by modern biologists. But it contains the kernels of thinking about how new species survive and accumulate through occupation of niches different from those of the ancestral species. Darwin was a remarkably original yet methodically encyclopedic thinker, who considered an even greater range of problems and phenomena than he is usually given credit for.

And there is yet another way in which Darwin retains a deep grip on modern evolutionary theory. As we'll see in some detail by following his conceptual path as he developed "my theory," Darwin came to emphasize certain lines of evidence over others, and certain processes over others. He waffled on the importance of geographic isolation. He was so convinced of the central power of natural selection that he rejected phenomena that didn't seem to fit in with his vision of how selection works through geological time, and so unwittingly bypassed legitimate fruits of the evolutionary process. There is to this day a prominent strain of "selfish gene" evolutionary thinking that essentially sees evolution as *only* a matter of natural selection working on the spectrum of heritable variation present in each generation. Isolation, speciation, the role of extinction—none of these larger-scale phenomena impinges to any significant degree on such thinking. This is all hauntingly similar to the version of evolution Darwin chose to present to the world when he finally acceded to the inevitable and broke down and wrote *On the Origin of Species*.

It is astonishing that one single person could still dictate so much of the agenda of evolutionary biology so many years after his death. No other historical figure in science retains such an influence. Who was this man? And how did he manage to accomplish what no others had been able to do before him—and accomplish it so well that he remains a powerful figure in the modern incarnation of the multi-faceted science he founded?

Charles Robert Darwin: A Thumbnail Sketch

Given the lasting importance of Darwin's seminal formulation of organic evolution, it is not surprising that a vast "Darwin industry" flourishes right up to the present day. Indeed, if anything, there have been more books and articles published on Darwin, his life and work, in the latter decades of the twentieth and early twenty-first centuries than ever before. Over 15,000 pieces of his correspondence have been collected—the vast majority housed at Cambridge University (see the Bibliography for a Web site address to this collection). And most of Darwin's early notebooks and manuscripts—as we shall see in greater detail in ensuing chapters—have now been deciphered, annotated, and published with meticulous care by a number of scholars.

First among these scholars was Darwin's seventh child and third son, Francis, who as a youth spent much of the year 1860 with his father when Darwin sought solace and distraction from the agony he felt after *On the Origin of Species* was finally published in November 1859. On the occasion of the centenary of Darwin's birth—simultaneously the fifty-year celebration of publication of the *Origin*—Francis published his father's *Sketch* of 1842, and the longer *Essay* of 1844, neither of which had previously seen the light of day. In 1963, Darwin's granddaughter Nora Barlow published some of his notebooks from the voyage of the *Beagle*, including what seems to have been Darwin's earliest written statement of his nascent evolutionary ideas.

The familial legacy persists in the more recent contributions of great-grandson Richard Keynes, who has catalogued Darwin's specimens and notes from the *Beagle*—and has written, as well, a general history of Darwin's voyage. Only when the British embryologist Sir Gavin De Beer made the first attempt to publish Darwin's "Transmutation Notebooks" did a member outside the family turn serious attention to the core of the writings—all previously unpublished—that preceded the *Origin*. The work has since gone on, with scholars and philosophers joining the Darwinian exegesis—long since (perhaps understandably) coming to outnumber similar efforts by biologists.

Other works abound, including imaginative novels and detailed examinations of specific aspects of Darwin's life and work. But perhaps the most important genre to have emerged is the "socially embedded" approach to biography, of which several outstanding examples have appeared in recent years. No idea in science has shaken society nearly so much as evolution—and in making that claim I include everything, even $e = mc^2$. Understanding the tone and temper of the times into which Darwin was born, and through which he lived his early life until he himself began to transform those times, is absolutely essential, especially if we are to understand why he waited over twenty years to publish his radical notions, as well as the genesis of those evolutionary ideas in the first place. Janet Browne's two-volume biography on Darwin, the equally impressive opus of Adrian Desmond and James Moore, and Randal Keynes's study are the very best of these socially and psychologically imbued biographies.

Yet there remains such a thing as pure intellectual history. Thanks especially to their superb scholarship and the ability to read Darwin's handwriting (no small feat in itself) of Sandra Herbert (the "Red Notebook") and David Kohn (the "Transmutation Notebooks"), we can now retrace Darwin's evolutionary thinking virtually from its inception. Darwin's conceptual odyssey with evolution is, to my mind, even more compelling than the many adventures that he experienced on his epic five-year voyage on the *Beagle*.

Difficult as it seems to find a niche not already exploited by the numerous writers in the industry who have come before me, I have nonetheless found in Darwin's early notebooks and manuscripts a story of intellectual creativity—growth and change, itself a form of personal evolution—that for the most part has not as yet been fully appreciated. My own perspective is as an evolutionary biologist, whose passion and central goal throughout my career has been to achieve a better fit between what we think we know about the history of life, on the one hand, and, on the other, the ideas we have about how evolution happens: the processes that conspire to create that history we study.

Most especially I am a paleontologist. Darwin was first and foremost a geologist, and it was his observations on South American fossils while traveling on the *Beagle* that helped convince him of evolution. Yet, by the time he published the *Origin*, Darwin was downplaying the importance of fossils and the other original patterns that had led him to the idea of evolution in the first place. We need to know why he chose to do so—for, as I have already remarked, his actions have effectively locked evolutionary biology on its course right down to the present day. This is my niche: an intellectual history of Darwin's evolutionary thinking by a passionate Darwinian who happens to be a practitioner of one of the many fields enveloped by evolutionary theory.

A final note on sources: I have never encountered anyone as clear on his own manner of thinking as Charles Robert Darwin. He was always transparently honest when he wrote, above all with himself. He was very explicit about how he thought about things, how and why he reached his conclusions, but also what from an analytic, philosophical point of view his method really was. This proves to be critical, as a close reading of Darwin reveals the creative process that led him to evolution in the first place, and allows us to trace as well how and why his thinking changed as time went along.

In 1876, Darwin wrote an *Autobiography* that, he says, was intended for the eyes of his family only. Again, there is no reason to doubt him on this; after all, so many of his manuscripts in his earlier years had been written for his own eyes only. Darwin, like anyone else, may have not been privy to every psychological truth about himself: the digestive misery he suffered so severely throughout his entire adult life, it is now commonly agreed, reflected more his anxiety over harboring his dark secret of evolution than anything systemically, medically wrong with him. But in all his writings, personal and scientific, he is as guileless as a writer can be. All this is to say that the *Autobiogra-*

phy itself is a welcome, trustworthy source on Darwin's life—one from which I quote throughout the following brief biographical sketch.

The Early Years. Charles Darwin was born into privilege, with wealth on both sides of the family. His two grandfathers, both of whom died before Charles was born, were well known in their own days. Erasmus Darwin was a medical doctor still remembered for his writings, often in verse form. His most lasting contribution—famous largely because of the work of his grandson on the same themes—was *Zoonomia*, an examination of physiology, health, and biological nature, including what we now call evolution. Charles's other grandfather, Josiah Wedgwood, founded the famous pottery works. The two men were friends, and among the emerging intelligentsia of late Georgian England. Together with such early scientific notables as Joseph Priestley and James Watt, they were members of the Lunar Society—a band of inquiring minds at the forefront of science, medicine, and technology that had the grace and good humor to refer to themselves as "Lunaticks."

The Wedgwoods and Darwins were formally united when Susannah, eldest child of Josiah and Sarah Wedgwood, married Robert Waring, third child of Erasmus and Mary Darwin. Charles Robert Darwin was the fifth of their six children; three daughters (Marianne, Caroline, and Susan) came first, then a boy (Erasmus), Charles, and then another sister, Catherine.

His mother died when Charles was only eight. Darwin grew up at "The Mount," their home in Shrewsbury, under the caring guidance of his three elder sisters, who basically took over, running the household in the absence of their mother. Charles's older brother Erasmus ("Ras") was also a key figure in those early days. His father, Dr. Robert Darwin, was benevolent but apparently rather remote. Charles was fortunate to have his older sisters and brother.

He was lucky, too, in having the Wedgwood clan living nearby. Uncle Josiah ("Jos") was a warm and welcoming figure in his life. Living at Maer Hall with his own brood of seven, Josiah Wedgwood was to prove a central figure in Darwin's life: first simply through the friendly atmosphere of his home; then, as time went on, as an adviser and mentor (it was Josiah's support that helped break down Darwin's father's resistance to letting Charles join Captain FitzRoy as naturalist on the *Beagle*). And seventh among Josiah's children was Emma, born in 1808, who was to marry Charles in 1839.

The Wedgwoods, including Darwin's mother and Emma, his wife-to-be, were Unitarians, thus "Nonconformists" in a genteel world where the vast majority were communicants of the Church of England. The Wedgwoods and the Darwins were in many ways liberals in terms of the political issues of the day (though Ras and Charles apparently toyed with right-wing views as students together, until scolded by their older sisters). Ras was to go on to host dinner parties in London which the young Charles attended regularly when he returned after his five-year-long voyage—dinner parties where guests such as Harriet Martineau were decidedly left of center in their political thinking.

Thus to some degree Darwin grew up in a family environment that was one of privilege but that emphasized social consciousness as well. The two are not mutually exclusive, but (as is still true today) wealth and a sense of social responsibility often do not go hand-in-hand, a circumstance that I think helps explain why Darwin felt compelled to follow his thinking to its logical conclusion, extending his evolutionary conclusions to include mankind virtually from the beginnings of his recorded thoughts on the subject. But at the same time he was reluctant to rock the boat, shock the world, and disrupt Victorian social order by revealing his ideas—which became his deep dark secret.

After Darwin's mother died, he was sent to a private school in Shrewsbury for the *de rigueur* classical education. But Charles vastly preferred the outdoors, and even in these early days was drawn to collecting natural history specimens, a passion he shared with Ras. Together they also conducted chemistry experiments in a shed on the property. When Darwin turned his attention more fully to the delights of shooting game (an activity that pre-adapted him well to the role of naturalist-collector on the *Beagle*), his father famously erupted (as Darwin himself recalled in his *Autobiography*): "You care for nothing but shooting, dogs and rat-catching, and you will be a disgrace to yourself and all your family."

So he shipped his younger son off to Edinburgh at age fifteen to join Ras in the ranks of medical school students—to carry on a family tradition, and above all else to gain a profession, a livelihood and identity that would carry him through adult life. But this early foray into gaining a profession and hence future respectability didn't take with young Charles. Nor did Ras end up practicing medicine. Both brothers remained close throughout the remainder of their lives (Ras died in 1881, only a year before Charles), but Ras never did develop any strong professional interests or identity.

Charles by contrast was to prove himself a very ambitious young man, but this was far from evident in his early days. One reason is blindingly obvious from yet another key remark in the *Autobiography*. Speaking of his time in Edinburgh, Darwin writes: "But soon after this period I became convinced from various small circumstances that my father would leave me property enough to subsist on with some comfort. . . ." With no worries about food and shelter, Darwin melded this deterrent to getting serious about finding work with a distaste for the practice of medicine itself. He was especially disturbed by the agonizing screams of surgery patients operated on without anesthesia. Talk about the need for a strong stomach: Darwin simply could not abide the cruelty and suffering of the medicine of his day.

Charles's experiences in Edinburgh did provide a crucial first step in his professional path. He kept up his passion for natural history through his beetle collecting (which had started back home in Shrewsbury), but something much more: for the first time he was exposed to the excitement and the rigors of scientific research and debate. He became a member of the Plinian Society, and delivered his first scientific paper, on a marine invertebrate species living in the Firth of Forth. Most important, he met the young Professor Robert Grant, an ardent follower of Lamarck's evolutionary thinking, and an admirer as well of Darwin's grandfather Erasmus's *Zoonomia*.

But Darwin's decided lack of enthusiasm for medicine soon became apparent to his bill-paying father, who decided after just eighteen months to place the boy at Cambridge to be trained as a clergyman, despite his mother's Unitarian disposition. As Dar-

Cartoon of Charles Darwin riding an enormous beetle, drawn by his friend Albert Way. Darwin's passion for beetle collecting developed early in life, and although beetles never figured prominently in his evolutionary notes, manuscripts, and books, nonetheless the riot of insect diversity in England clearly inspired the young Darwin to observe, collect, and think about the living world.

win himself wrote in his *Autobiography*, "Considering how fiercely I have been attacked by the orthodox, it seems ludicrous that I once intended to be a clergyman."

Darwin arrived in Cambridge in late 1828—graduating in 1831, if not with honors, at least tenth in the list of non-honor students. Cambridge honed Darwin's taste for the informal side of natural history: the pure joy of collecting plants and especially his beloved beetles—a hobby that was to blend so beautifully with his serious intellectual pursuits for the remainder of his long life. He writes in his *Autobiography* that "no pursuit at Cambridge was followed with nearly so much eagerness or gave me so much pleasure as collecting beetles." And it was at Cambridge that this other side, begun at Edinburgh, started to take shape. The botanist and geologist John Stevens Henslow introduced Darwin to the professional side of natural history, especially botany. Darwin became the "man who walks with Henslow"—and their relationship was soon to pay enormous dividends.

Henslow was a clergyman, as was another Cambridge faculty member, Adam Sedgwick. Sedgwick was one of the early important geologists in Great Britain, and the week-long summer excursion to Wales in August 1831 that Darwin undertook with Sedgwick (at Henslow's suggestion) amounted to some of the purest field training in any subject that Darwin ever received. Darwin still referred to himself as first and foremost a geologist throughout the rest of his life; indeed, his initial reputation in science came through his excellent geological work while on the *Beagle*—some of which was published by Henslow under Darwin's name, based on the letters he received from the young naturalist still out on the voyage.

The Voyage of the* Beagle*, 1831–36. When Darwin got back from Wales, he received the fateful letter from Henslow that was to plunge him into what he called "the most important event of my life." As Darwin himself recalls in his *Autobiography*: "On returning home from my short geological tour in North Wales, I found a letter from Henslow, informing me that Captain Fitz-Roy was willing to give up part of his own cabin to any young man who would volunteer to go with him as naturalist to the Voyage of the *Beagle*." Darwin's father was opposed; but, as Darwin said, fortunately his father had added that "if you can find any man of common sense who advises you to go I will give my consent." So Darwin initially turned down the offer. But Uncle Josiah Wedgwood intervened, Darwin's father relented, and Darwin was free to go.

The trip lasted nearly five years, from December 27, 1831, to October 2, 1836. Darwin had been eager to visit the tropics, and had read von Humboldt's account of his voyage to the Canary Islands and South America—even before the invitation to join the *Beagle* he had been scheming about ways to visit Tenerife. He was as eager and prepared as a young man could have been to make the most of this opportunity. Throughout the voyage (and despite severe seasickness that made his long explorations on shore all the more welcome), Darwin enjoyed the reputation of being one of the two or three most energetic men on board.

Captain Robert FitzRoy had, Darwin later wrote, "a most unfortunate" temper. Indeed, their famous argument over slavery while in Brazil cost Darwin his dining privileges with the captain—until FitzRoy himself apologized. Darwin had witnessed the slave market in Bahia, Brazil, and had been horrified by the brutality he saw, and by the sheer inhumanity of separating husbands from wives, and parents from their children. Though he was to remain throughout his life typically Victorian in his elitist views toward women and "savages," in another sense he also keenly felt the essential humanity of all the people he encountered on the voyage—be they "savages," slaves, or colonists. What he could not abide was bad behavior.

The commission of the *Beagle* was to chart the coastal waters of southern South America. On an earlier voyage, FitzRoy had brought home to England several Fuegian people—the famous naked savages of the wilds of Patagonia. They had learned to speak English and readily caught on to the basics of British social behavior. Now three of them were being returned to their tribe. FitzRoy was appalled when a year later he found one of them, once again naked and with hair matted, no trace of Victorian niceties to be seen. It was a failed experiment in FitzRoy's mind, but to Darwin an unforgettable lesson in the behavioral malleability of men. To Darwin, it was a clear example of how superficial the potentially transient differences between gentleman and naked savage can be. On the surface, there seems to be all the difference in the world; but underneath lies an indication of the common experience of *Homo sapiens*, no matter the state of "civilization."

Darwin began to find fossils and observe large animals such as rheas (South American "ostriches") and guanacos, vicuñas and alpacas—observations that were to lead (along with his experiences later in the Galapagos) to his notion of evolution, arguably before he even reached home at the end of the lengthy voyage. These are details for the ensuing chapters. But Darwin's collecting was comprehensive, and his geological observations imaginative, detailed, and voluminous. He began sending specimens home—along with letters setting forth many of his geological results. He had a copy of

Volume 1 of Sir Charles Lyell's *Principles of Geology* with him—as we shall see, a key component in the emergence of the idea of evolution in Darwin's fertile and very open mind. But Lyell's book was also a further impetus to his own "geologising" as Darwin was to provide incontrovertible evidence that the Andes had risen (through a long series of earthquakes, one of which he experienced in 1835 in Valdivia, Chile) from beneath the ocean's waves. His geological cross section of the Andes is a classic rendering of the temporarily static results of deep-seated, intense, and episodically constant violent wrenchings of sediments quietly deposited on the seafloor and transformed into the slopes of prodigious mountains.

Later in the voyage, Darwin studied coral reefs, making important observations suggesting a revision in existing theory on the formation of coral atolls. (His results were published as *The Structure and Distribution of Coral Reefs*, 1842.) He found and collected many new species, a number of which were eventually to be named after him. In one of

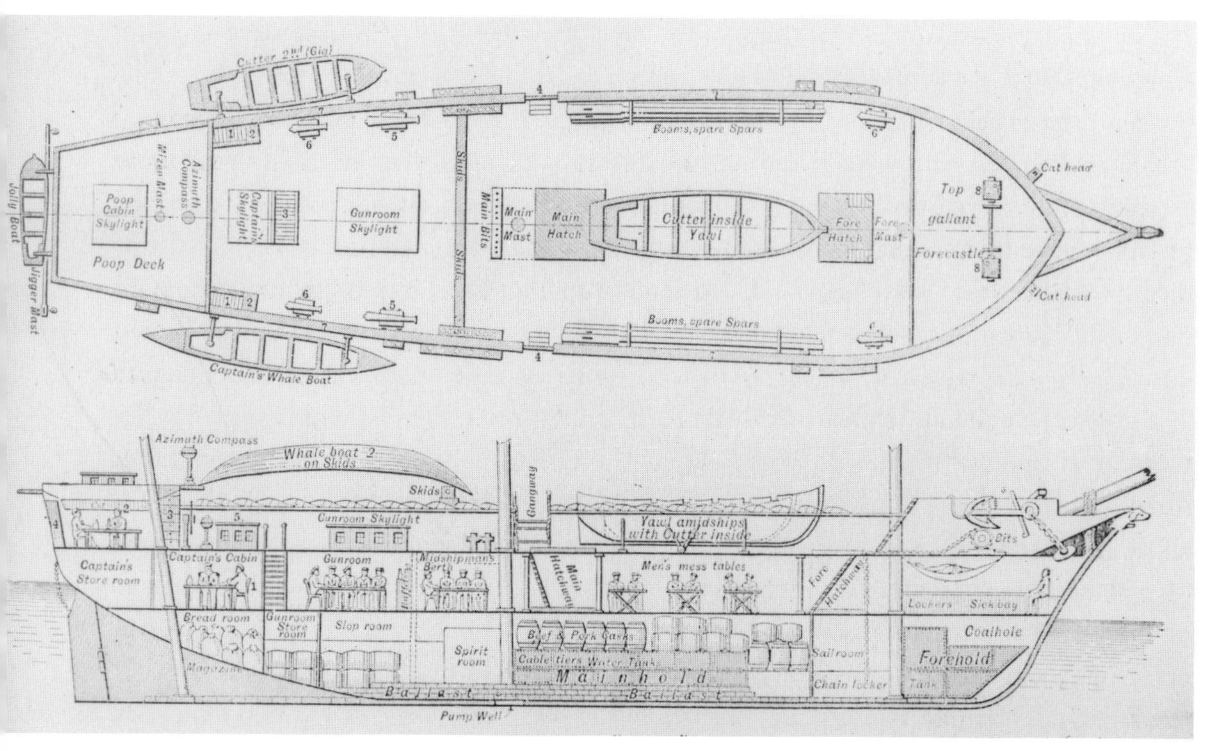

Drawings of HMS *Beagle*. The *Beagle* was a ten-gun brig, built eleven years before the epochal voyage that bears her name. She was only ninety feet long and twenty-four feet at the widest place.

The beached *Beagle* along the Santa Cruz River; the vessel was beached to undergo repairs. Conrad Martens.

his characteristically self-contradictory statements, Darwin writes in his *Autobiography*: "As far as I can judge of myself, I worked to the utmost during the voyage from the mere pleasure of investigation, and from my strong desire to add a few facts to the great mass of facts in Natural Science. But I was also ambitious to take a fair place among scientific men—whether more ambitious or less so than most of my fellow-workers, I can form no opinion." Darwin's hunt for fame, and his desire to avoid having to become a clergyman on his return home, were an energizing part of his zeal for hard work, and played no small role in his need to develop something radical: his theory of evolution.

The crew of the *Beagle* took their time charting the waters of southern South America, providing Darwin the much-needed opportunity for long stays on shore for his collecting and geologizing. They finally reached the Pacific Ocean (via the Strait of Magellan) in June 1834, and the Galapagos Islands in September 1835. They departed after five weeks, heading for Tahiti, New Zealand, Australia, and then Cape Town. Leaving South Africa, Captain FitzRoy decided to cross the Atlantic one more time, much to the dismay of Darwin and all the other crew members, who naturally thought they were homeward bound. Finally, FitzRoy relented and they made it back to England in October 1836.

Emma Wedgwood Darwin in 1840, shortly after her marriage to Charles. Emma was troubled when Charles (against his father's advice) compulsively told her of his work and growing religious doubts. She was especially concerned that they might not meet in the afterlife. But she remained his lifelong supporter until Darwin's death in 1882.

Charles Robert Darwin in 1840. Though he had not yet written an essay developing a coherent description of his evolutionary views, the Darwin shown here had already developed all the components of his "excellent theory"—evolution through natural selection.

London: Marriage and Revolutionary Thinking, *1837–42.* If the nearly five years spent on the *Beagle* gave Darwin the direction and mental ammunition for what was to prove to be his true life's work, it was during the five years after his return to England that he developed his theory of evolution, became firmly established in the scientific world, married Emma Wedgwood, and had the first two of their ten children.

Darwin returned home to find himself already known in scientific circles, thanks to his older mentors and friends, including especially Henslow and Lyell. Though he journeyed often to Maer Hall and The Mount in Shropshire, and visited Cambridge as well, Darwin lived in London—moving to a house on Upper Gower Street after marrying his cousin Emma in 1839. He became active in the Geological Society (against his will agreeing to become its secretary in 1838) and the Athenaeum Club. In January 1837, shortly after arriving home, he delivered a paper on the elevation of the coastline of Chile to the Geological Society.

He also busied himself with writing up his first book, published as part of a series of works on the results of the voyage under the general editorship of Captain FitzRoy: *Journal of Researches into the Geology and Natural History of the Various Countries Visited by H.M.S.* Beagle, *under the Command of Captain FitzRoy, R.N. from 1832* [sic] *to 1836.* Surviving in print to the present day, the book of course is more familiarly known simply as *The Voyage of the* Beagle. It was popular in Darwin's day—much to his delight, even though he apparently never realized a penny in profit from its sale. He later did profit from this experience in another way, probably becoming one of the very first authors to demand an advance against royalties when years later he submitted the manuscript of the *Origin* to the publisher John Murray.

Then, too, there was the matter of the proper scientific study of all of the specimens Darwin had collected on the voyage: rocks, minerals, fossils, marine invertebrates, terrestrial animals, and plants. He was able to enlist the aid of the ornithologist John Gould to study his birds fairly promptly, but it was not until the late 1830s and early 1840s that the results of the study of all Darwin's collections really began to see the light of day.

But the real event in Darwin's professional life—the one that instantly induced so much anxiety that he once again fell prey to stomach complaints and heart palpitations (as earlier before leaving on the *Beagle*)—was hidden from view: "In July [1837] I opened my first notebook for facts in relation to the *Origin of Species*, about which I had long reflected, and never ceased working for the next twenty years." We'll examine the contents of these notebooks—and the two essays written in 1842 and 1844 (the latter after he moved to Down House)—in some detail in chapters 3 and 4, for they record the development of the very core of Darwin's theory. These early documents provide an extremely rare glimpse into the very nature of the creative process.

Darwin realized early on that men were no exception to his thesis that all species on earth are related through a process of descent. As he later wrote in the *Autobiography*: "As soon as I had become, in the year 1837 or 1838, convinced that species are mutable productions, I could not avoid the belief that man must come under the same law. Accordingly I collected notes on the subject for my own satisfaction, and nor for a long time with any intention of publishing." Some of these notes are on the orangutan

Jenny the orang—one of the two "Jennys" who were the first orangutans to come to the Regent's Park Zoo in London. Darwin found them fascinating, and compared their behavior with that of his first two children, William and Anne. Queen Victoria pronounced the second Jenny "disagreeably human."

Jenny at Regent's Park Zoo, comparing Jenny's behavior with that of his first two children, William (born 1839) and Anne (born in 1841). Darwin loved his children dearly but saw no conflict in comparing their behavior as infants with that of a great ape. Tradition, of course, had it that any resemblance between man and ape was purely coincidental—man having been created by God in His image.

***Down House**, 1842–82.* Almost as if to say that the last forty years of his life were an uneventful afterthought, Darwin writes in his *Autobiography*:

> Few persons can have lived a more retired life than we have done. Besides short visits to the houses of relations, and occasionally to the seaside or elsewhere, we have gone nowhere. . . . I have therefore nothing to record during the rest of my life, except the publication of my several books.

And it is certainly true that the quiet, domestic life that Darwin led in Kent represents a retreat from the world that would not have been predicted from his younger

In a dramatic testimony to his continuing influence in the twenty-first century, Charles Darwin replaced his fellow Victorian Charles Dickens on the British ten-pound note in 2000.

John Herschel, a prominent scientist and philosopher during Darwin's formative years. Darwin visited Herschel in Cape Town toward the end of the *Beagle* voyage, possibly discussing the "mystery of mysteries" as well as geological topics.

Erasmus Darwin, Charles's grandfather and author of *Zoonomia*, a lengthy work that featured organic evolution among its themes.

Alfred Russel Wallace, the naturalist who, in part inspired by Darwin's *Voyage of the* Beagle, began his explorations in the 1840s already convinced of evolution. His 1858 manuscript describing natural selection prompted their joint publication, and galvanized Darwin into writing *On the Origin of Species*.

Francis Galton, Darwin's cousin and a founder of eugenics. Charles Darwin had virtually nothing to do with the various aspects of "social Darwinism" promulgated under his name.

Josiah Wedgwood, Darwin's other grandfather and founder of the famous pottery works. Josiah Wedgwood and Erasmus Darwin were friends and fellow members ("Lunaticks") of the Lunar Society.

Susannah Wedgwood, daughter of Josiah and mother of Charles Robert Darwin. Susannah died when Charles was eight; he could hardly remember her later in life.

Dr. Robert Waring Darwin, Charles Darwin's father. Though sometimes despairing of his younger son's future, Charles's father was basically supportive and proud of his accomplishments.

Charles at age seven, with his younger sister, Catherine. Darwin's older sisters effectively raised Charles, Ras, and Catherine after the death of their mother.

The Reverend John Stevens Henslow, botanist and geologist. Darwin became the "man who walks with Henslow" during his days at Cambridge. Henslow taught Darwin to be meticulous in his collecting and note-taking; facilitated his geological training with Reverend Sedgwick; got him the position as ship's naturalist and companion to FitzRoy on the *Beagle*; received Darwin's shipments of specimens while on the voyage; and published some of his geological observations—establishing Darwin's reputation as a scientist before he reached home in 1836.

The Reverend Adam Sedgwick, an early important geologist at Cambridge. Darwin considered himself primarily a geologist throughout his adult life. The field excursion to Wales with Sedgwick in the summer of 1831 was an especially important part of Darwin's early training as a scientist.

August 31

My dear Father
I am afraid I am going to
make you again very uncomfortable.—
But upon consideration, I think you will
excuse me once again stating my
opinions on the offer of the Voyage.—
My excuse & reason is, is the different
way all the Wedgwoods view the subject
from what you & my sisters do.—
I have given Uncle Jos, what I fervently
trust is an accurate & full list of your
objections, & he is kind enough to give his
opinion on all.— The list & his answers
will be enclosed.— But may I beg
of you one favor. it will be doing me

Portion of Darwin's letter to his father, including a list (facing page) of his father's objections to his joining the *Beagle* as ship's naturalist. It was his uncle Josiah Wedgwood who convinced Darwin's father to relent and allow Charles to accept the offer and take the epochal journey.

10

(1) Disreputable to my character as a Clergyman hereafter

(2) A wild scheme

(3) That they must have offered to many others before me, the place of Naturalist

(4) And from its not being accepted there must be some serious objection to the vessel or expedition

(5) That I should never settle down to a steady life hereafter

(6) That my accomodations would be most uncomfortable

(7. That you should consider it as again changing my profession

(8) That it would be a useless undertaking

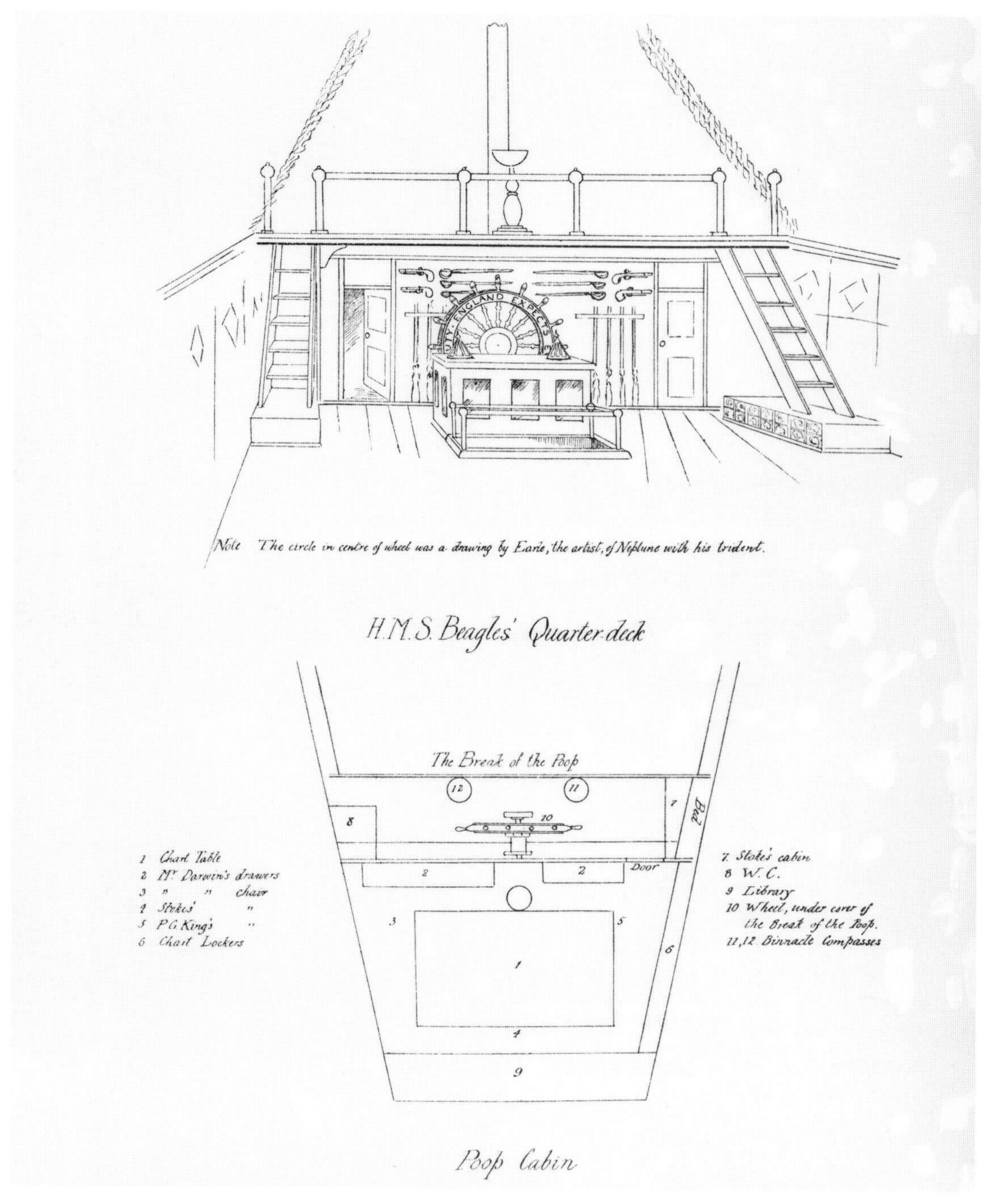

Poop deck of the *Beagle* with Darwin's cabin. Darwin had to stoop to get through his cabin door. By day he shared the tiny cabin (10' by 11') with a shipmate, studying his specimens and writing his notes; at night, he slung his hammock over the table.

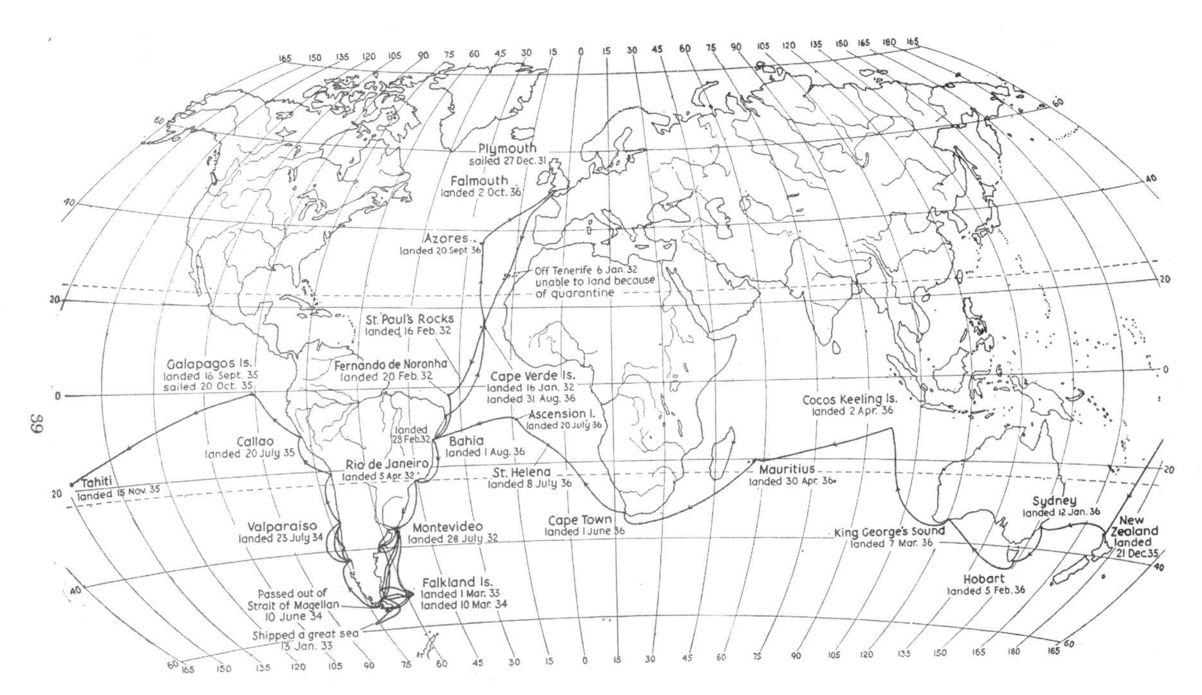

Map of the *Beagle*'s five-year voyage. The primary purpose of the trip had been to chart the waters of the southerly coasts of South America.

Captain Robert FitzRoy, commander of HMS *Beagle*. FitzRoy's conservative views led to some famous flare-ups with Darwin, though on the whole the two got on well and respected one another. FitzRoy was appalled, however, when Darwin published his evolutionary views decades later.

Crossing the equator on the *Beagle*. Darwin was subjected to the usual ritual dunking that everyone crossing the equator for the first time had to suffer through. Original by Augustus Earle, along with Conrad Martens, one of the two illustrators on the voyage.

days. Trains had just recently become part of British life, and the quiet little village of Downe southeast of London no longer seemed so removed from the life he had been leading. The Darwins felt that a move to the quieter, less polluted countryside would provide more room for their growing family and the sort of ambience both recognized from their youth—an environment conducive to quiet contemplation and writing.

Darwin's virtually daily bouts with ill-health incapacitated him, keeping him from socializing ("I have therefore been compelled for many years to give up all dinner parties; and this has been somewhat a deprivation to me, as such parties always put me into high spirits"). Yet in another section of his *Autobiography* he writes: "Even ill-health, though it has annihilated several years of my life, has saved me from the distractions of society and amusement"—more or less letting the cat out of the bag. Darwin increasingly shunned society, and when he traveled, it was only to the safe houses of relatives or the closest of friends, or to take the curative waters of Malvern.

As we shall see in the next chapter, Darwin was a highly intuitive, creative thinker. But he was also an experimentalist, methodically testing his ideas in his greenhouse and elsewhere in his gardens. He continued his close observation of nature around him: in his book *Annie's Box*, Randal Keynes recounts the story, passed down in the Darwin family, that Jessie Brodie, the children's nurse and a former employee of William Makepeace Thackeray, once said, after seeing Darwin staring at an anthill for an hour, that it was a pity that Mr. Darwin, unlike Mr. Thackeray, had nothing to do. That anecdote seems to capture the tone and tenor of Darwin's daily life at Down perfectly.

Details of Darwin's scientific life—and especially his thinking and writing on evolution—will emerge in the next chapters. Suffice to say here that much of Darwin's scientific work at Down House in fact does not appear in *On the Origin of Species* (1859) or in *The Descent of Man* (1871). He spent eight years (two of which, he says, were actually lost to illness) working on barnacles, beginning in 1846. Though it is sometimes said that this careful monographic study on the "systematics" (classification) of barnacles was crucial to the development of his evolutionary ideas, those ideas were in fact already in place in his mind—and in his *Essay* of 1844. Rather, the barnacles gave Darwin something to do, providing experience, true enough, in the practical field of biological classification, but serving really as a distraction from what he might have otherwise been doing: writing up his evolutionary ideas for publication. Though publication of his barnacle work did embellish Darwin's scientific reputation, barnacles hardly figure in the *Origin* at all.

But pigeons do. After Annie's death in 1851, within a few years Darwin began writing up what he planned to be his masterwork—to be entitled *Natural Selection*. He also decided to become expert in some area of selective breeding and chose to become deeply involved with the entire subculture of fancy pigeon breeding. When circumstances forced him to abandon *Natural Selection* and to publish instead the shorter *Origin of Species*, his experimental work with pigeons played a prominent, early role in the presentation of his evolutionary ideas. Indeed, one pre-publication reviewer of the *Origin* engaged by the publisher John Murray thought the only part of the manuscript worth publishing was Darwin's account of pigeon breeding!

After the *Origin* was published in 1859, Darwin plunged afresh into botanical work—both in the field and experimentally. The work has significance in the history of botany, but again seems to have been conducted, however enthusiastically, as much to fill time and distract him from the reactions of readers of the *Origin* as to solve further problems in heredity and variation. Darwin's later work on earthworms, insectivorous plants, climbing plants—all of it is worthy. And though he was to do little practical work in geology after he moved to Down House, had he not come up with his evolutionary ideas, if Darwin were to be remembered at all today, it would be for his geological work—both in the field, and his later arguments on the age of the earth.

But the truth of the matter is that Darwin is remembered today—indeed, cannot be forgotten—because of his ideas on evolution. He was a true revolutionary, ironically enough, buried at Westminster Abbey after his death in 1882 in a hurried attempt to incorporate him into the pantheon of the greatest of British figures. He was a reluctant revolutionary. His curiosity and frankly ambitious desire to "take a fair place among scientific men" led him to evolution—and the awful prospect of going against the unspoken yet no less mighty strictures of British society. Emma once said that she knew of no more honest or open a man ("transparent" was the word she used). As we'll now see, he was at his most thoroughly and openly honest when he was writing for his eyes only. But he maintained that deep integrity when finally forced out into the open world of publication. All of this—the man, his mode of thinking, and his ideas themselves—demands a much closer look.

CHAPTER 2

Darwin on the Sandwalk

Imagine, for a moment, Charles Darwin taking one of his daily walks along his beloved "Sandwalk"—that stretch of Kentish gravel Darwin had built along the rear edge of his property at Down House. Outwardly, things were normal. It is mid-June in 1858. Darwin picks up his walking stick and goes out the back door of Down House. He is a middle-aged man of forty-nine years—a man of regular habits. Strolling along the Sandwalk, a path lined with bits of chalk, flint, and the occasional fossil from the local Cretaceous bedrock of southeastern England, he regularly took the air, inspected his grounds—and mulled over his life and work. He often found peace walking along the path that lay between his expansive field and his neighbor's adjoining property. And he occasionally felt that exhilaration that comes when sudden insight pops into the brain—a solution to a nagging problem that often did not even seem to be uppermost in one's mind at the moment. Darwin was a highly intuitive man, a man whose capacity for creative thought was perfectly matched by his rigor in testing his ideas with observation and analysis, *de rigueur* in the freshly minted practice of modern professional science.

But things were far from normal that day. For one thing, his two-year-old retarded child, Charles Waring, was battling scarlet fever and near death. His fifteen-year-old daughter Etty (Henrietta) had diphtheria, and Darwin had already lost his dear Annie seven years earlier. She had been the second child, Emma and Charles's first daughter;

her death from tuberculosis in 1851 had removed what little was left of his religious faith. All this was enough to make him upset. Though another son, Francis, was to write many years later that his father hardly knew a day when his health was robust and normal, that June day his stomach must have been in even greater turmoil than usual.

But there was more, far more, upsetting Darwin's stomach that day. For Charles Robert Darwin had lived the past twenty-two years as a man with a secret. And his was not just any old garden-variety secret like a clandestine love affair, or commission of a crime (though he thought of it in those terms). He had traveled the globe as a young man in the 1830s, gentleman ship's naturalist and companion to the *Beagle*'s captain Robert FitzRoy, and by the time he reached home in late 1836, not quite twenty-eight years old, he was convinced that life had evolved through natural causes. He saw that human beings were no exception: we are part of the spectrum of life along with all other species of animals, plants, and the then largely unknown microbial world.

And he had told almost no one about it. True, he had told his father that his work was forcing him to entertain grave doubts on matters of religious faith, and his father, Dr. Robert Waring Darwin, had sagely advised him against telling his fiancée—advice the honest Charles proceeded to ignore. Emma was concerned, but not because religious doubts were all that unusual and heinous (religious skepticism was rife in certain quarters in England, and had loomed in the family at least since the days of their mutual grandfather, Josiah Wedgwood, and Darwin's other grandfather, Erasmus). Rather, it was the thought that they might not be together throughout eternity that troubled Emma most about Charles's straying from the fold of Christian orthodoxy—a reaction that might seem quaint in the twenty-first century, but was very important to people living in the early nineteenth century who, after all, were all too aware of the shortness of life. Medicine was still primitive, antibiotics were far in the future, and people simply did not take a long life for granted the way we tend to do today.

Keeping his secret to himself, Darwin told no one *why* he had begun to have doubts, though shortly after their marriage in 1839, Emma certainly knew. He had even hidden away the first truly publishable manuscript on evolution—his 231-page *Essay* of 1844—with a note to Emma instructing her to publish it in the event of his death. He specified a sum of £400 to support publication, and suggested suitable editors for the work. Indeed, Emma had read the manuscript, pointing out to Charles the passages she found difficult to follow.

But otherwise, except for a few hints here and there to his friends, Darwin was quiet about his evolutionary ideas until he wrote his new friend, the botanist Joseph Hooker, about his doubts on the "stability of species" early in 1844—in a letter famous for his aside: "it's like confessing a murder."

And he had told Charles Lyell as well. In 1836, before Darwin's crucial meeting with John Herschel in Cape Town, Herschel, already considered an important British scientist, had written Sir Charles (a near-equal force in science in those days) a letter in which he wondered when someone would come forth with a naturalistic explanation of what Herschel called that "mystery of mysteries": the appearance of new species to replace those lost to extinction. Lyell was a good choice, as England's most accomplished and famous geologist and a man well versed in the fossil record. Yet in 1858 Lyell still maintained his belief in the fixity of species, a position he didn't abandon until some years after his young friend and colleague published *On the Origin of Species* (still very much in the future). Lyell remained a creationist largely because he feared the social and moral implications of evolution, especially the evolution of mankind.

Sir Charles Lyell, barrister and geologist, who expanded Hutton's uniformitarian views in his three-volume *Principles of Geology*. Darwin read all three on the *Beagle* voyage, and despite Henslow's admonition not to believe the contents of Volume 1, and Lyell's strong stand against evolution in Volume 2, was inspired in part by Lyell to develop a science of the history of life similar in many respects to Lyell's theory of earth history. Lyell came to accept evolution only some years after the publication of Darwin's *Origin* in 1859.

But there was more than religion lurking behind Darwin's fear of a backlash should he decide to come out of the closet and publish his ideas. In 1844, Robert Chambers, a writer and publisher in Edinburgh, pseudonymously published *Vestiges of the Natural History of Creation*, a pro-evolutionary broadside that had Britain agog. Orthodox scientists were in the vanguard lambasting the book—and Darwin saw to his horror the ridicule heaped on Chambers's arguments and evidence. Chambers had not done his homework as thoroughly as he might, and the lesson was not lost on Darwin. Anyone daring to challenge the fixity of species—that notion that species remain the same today as they had been when originally made by the Creator—had better have all arguments tightly formulated, with the solid, incontrovertible evidence laid out in no uncertain terms. The bright side was that Chambers had not scooped him; the dark side was that Darwin saw that his scientific colleagues in 1844 were still prepared to mount a full-scale attack on anyone daring to suggest that life had evolved through natural processes.

So he had bided his time—though he did hedge his bets in writing his summary *Essay* in 1844. This was an expansion of the 1842 *Sketch*, thirty-five handwritten pages so replete with notes to himself as not to be a truly polished piece of writing by any means, though its creative freshness and intensity makes this *Sketch* by far my favorite piece of Darwin's writings on evolution. A distillation of the notebooks written in the late 1830s right after the *Beagle* voyage, the 1842 *Sketch* captures much of the excitement of Darwin's creative, fertile mind at work—otherwise only seen in maddeningly elliptical snippets in his notebooks. Darwin was a man who kept his thoughts so much to himself that even the notebooks and early essays keep veiled the depths of his thinking. Soon after he started recording his evolutionary thoughts in his notebooks, Darwin was already calling evolution "my theory"—very much as if privately, so privately that he did not ever set down the full range of his thoughts on paper in those early years, he had the evolutionary story pretty much figured out long before he wrote anything down. I am with those historians who think that he had become a convinced evolutionist even before he set foot back on British soil at the end of the *Beagle*'s journey in late 1836.

But something was still missing from the theory. Many years after the publication of *On the Origin of Species by Means of Natural Selection* in 1859, Darwin wrote that he had early on realized that a mere recounting of facts of natural history—the patterns of variation in space and time of plants and animals that all point to the conclusion that all species on earth are related by a process of common descent (and, as he put it, "with modification")—wouldn't be enough to convince the world that life has evolved. True, the patterns he saw while traveling down the length of South America, examining geographic distributions of modern species, and fossil remains, and when he got to the Galapagos were enough to convince him of the fact of evolution. But he thought that, to make an ironclad case for evolution, he needed to explain *how* evolution happens.

And he had found that mechanism less than two years after arriving home. He had read Thomas Malthus, and realized that more organisms of any given species are born each generation than are needed simply to replace their parents. He began to see a process in the wild analogous to the actions of agricultural breeders, who consciously "select" for the properties they wish to see embellished in their domestic varieties of flowers, pigeons, sheep, and so forth. Realizing that, whatever the cause, offspring tend to inherit from their parents (and grandparents) certain features, he put it all together in his principle of "natural selection": that of all offspring produced each generation in the wild, only those best suited to surviving and eking out a living will, on average, be the ones who will successfully breed and themselves leave offspring to the next generation. The characteristics that gave their parents a slight edge in what he called the "struggle for existence" would tend to be present in their offspring, then passed along again—until such time as conditions changed and other variants might have an edge in that struggle for existence.

Voilà: "natural selection," very sketchily developed, but definitely there, in his "Transmutation Notebooks" D and E (written in 1838–39), and fully developed, specifically named, in his 1842 *Sketch*. By the time the *Sketch* was written, it was pretty clear that Darwin had come to see natural selection itself as "my theory"—rather than the concatenation of repeated patterns of the geographic and geological distribution and variation of species, patterns that made clear to Darwin in the first place the *fact* of evolution.

But instead of publishing his theory, Darwin monographed barnacles, contributing hardly at all to the ideas finally expressed in the *Origin* in 1859. He had narrowly escaped being preempted by Chambers in 1844—and learned a valuable lesson about the absolute necessity of thoroughness and rigor in the eventual public presentation of his views.

Darwin had finally begun writing his magnum opus, to be entitled *Natural Selection*, on May 14, 1856. As he trudged along the Sandwalk that June day in 1858, he had already amassed some ten chapters, and there was still a long way to go. He was cramming in virtually all the examples he had found in his life that supported his ideas: observations he had made as long ago as the *Beagle* voyage in the 1830s; analyses by experts of some of the plants, animals, and fossils he had brought back from that epic voyage; but also assorted facts on natural history that he had amassed over the past twenty years from the newly emerging scientific literature—and not least, from the dozens of contacts he had made through correspondence with botanists, zoologists, geologists, and plant and animal breeders throughout the world.

His correspondents included a young man named Wallace who, in part modeling himself after Darwin, had been studying the fauna and flora of the far-flung islands of what are now Malaysia and Indonesia. Darwin had written Wallace, encouraging him to pursue his work, including his thoughts on species.

But, prodigious as this fledgling manuscript on *Natural Selection* had already become, it was too little, too late. What was bothering Darwin the most that June day in 1858 was the arrival, a few days before, of yet another letter along with a manuscript from that far-off naturalist and specimen collector, Alfred Russel Wallace. With it came a fresh round in Darwin's own personal "struggle for existence." For Wallace had truly scooped him—outlining a theory of natural selection (though he didn't call it that) so well that Darwin later said he could hardly have written a better abstract of his ideas himself.

Darwin was aghast. Though he was many years later to write in his *Autobiography* that he "cared very little whether men attributed most originality to me or Wallace," nothing in fact could have been further from the truth. Beneath his diffident demeanor lay the heart and soul of an ambitious man, one who had long since come to see science as his bailiwick and key to success in life. When Darwin was born, there were few or any paid scientists of any sort whatsoever. Western science was still pretty much in its infancy, and was the pursuit of men (as yet very few women were included) who had other means of support. This generally meant one of two things: men of independent means, usually translating to inherited wealth. Or members of the clergy, who had ample time to consider the natural world in addition to the duties of tending their flocks. Indeed, among the best and brightest of these clergy naturalists were men with something more: positions as professors at England's universities. The Reverend Adam Sedgwick, proctor at Cambridge, was one such early mentor, as was John Stevens Henslow, also a clergyman.

Sir Charles Lyell was no clergyman, but rather a barrister from the monied upper crust. Lyell was as aggressively hardworking as any paid scientist. He had taken and extended the work of James Hutton, a Scottish gentleman farmer and physician of the preceding generation. It was Hutton who first articulated the concept of "uniformitarianism," that the "present is the key to the past" (in the salubrious phrase of William Playfair, who presented Hutton's views in vastly more readable and digestible form than Hutton ever managed to do). Lyell had adopted this approach, which basically says that the laws operating in the universe now have always been there, so that the past can be interpreted by asking how these very same laws might shape the earth over long periods of time.

Hutton came along at a time when the dominant view in geology was cataclysmal: the biblical story of the Flood, for example, fit in rather well with an overall perspective that saw the earth—and its living inhabitants—undergoing a series of catastrophic events, with sudden upheavals casting up mountains and otherwise periodically wreaking havoc over the globe. Lyell was especially persuasive to the young Darwin that the application of simple processes—such as gentle erosion played out over the long expanse of geological time—was sufficient to account for most of the changes the earth had undergone throughout its history. Lyell was a "gradualist" rather than a "catastrophist."

But there was also an emerging class of young scientists who were neither wealthy nor given to clerical leanings. Indeed, they were often strongly opposed to what they saw as the stranglehold of religion on the rational contemplation of the natural world. These were the first paid professionals—like Sedgwick and Henslow, college professors, but unlike these men, definitely not clergymen. Nor did they teach at universities with a strong Church of England affiliation.

The first of these "modern" scientists that Darwin met was Robert Grant, a lecturer at the University of Edinburgh. Grant was the invertebrate zoologist who gave Darwin his initial taste of scientific training—collecting, analyzing, and writing papers on the invertebrates of the Firth of Forth. But there was something else about Grant: he was a convinced evolutionist. Grant was an ardent admirer both of Darwin's grandfather Erasmus's *Zoonomia* and of the eighteenth-century French zoologist Jean-Baptiste, Chevalier de Lamarck. Lamarck saw the skein of similarity that ran through the animal kingdom, and even seemingly linked plants up with animals. He saw life as a grand Chain of Being formed by a process of evolution that continually produced higher, more complex forms of life from lower forms.

Grant was convinced that there were some forms of simple, primitive marine life that were indeed true connections—direct links—between the animal and plant kingdoms. He thought that bryozoans ("moss animals") provided such a link; bryozoans are now understood to be colonial and rather advanced forms of animal life, and thus not the link between plants and animals Grant thought them to be. Darwin was drawn to these lower forms of marine life throughout the *Beagle* voyage, and in fact it was his dissection of a remarkably tiny barnacle that led him on an eight-year sidetracking odyssey to survey and monograph all the barnacles of the world.

Grant had moved to University College in London in the 1830s. Darwin had avoided him, as Grant's radical views on evolution were allied with a political philosophy that was unsavory to Darwin's monied and clerical peers in the scientific world. But Darwin had long since taken up with other young scientists of the salaried ranks—most notably Joseph Hooker, who had joined his father Sir William as a botanist at Kew Gardens. It was Hooker to whom Darwin had confided his thoughts on evolution being "like confessing a murder" in 1844—a man with whom he corresponded continuously, and the one whose opinion on all manner of things counted perhaps the most to Darwin. Hooker, like Lyell, had never come to agree with Darwin on evolution up to that day in June 1858; as Darwin himself wrote in his *Autobiography*, "even Lyell and Hooker, though they would listen with interest to me, never seemed to agree."

There were, increasingly, other salaried professionals. Most notable, in Darwin's life at least, was Thomas Henry Huxley, already established as a professor at London's School of Mines. Huxley was later to go on to play a major role as "Darwin's bulldog"—appearing in public, taking on all comers as the leading evolutionary protagonist in the great fight unleashed by the publication of Darwin's *Origin of Species*. Darwin did know Huxley already, as a young, ambitious anatomist who loved to take on the dominant figure of the anatomical profession of his day, Richard Owen, superintendent of the Natural History Collections at the British Museum.

Owen was steadfastly opposed to evolution, which was a popularly held view in many medical establishments. Anatomists, though, tend to be duly impressed with the intricate complexities of organic structures, and it is often difficult for them to see how something as complex as, say, the vertebrate eye can evolve from simpler structures. Owen believed, instead, in "archetypes"—basic anatomical patterns laid down by the Creator, with different species developing different permutations and combinations of the fundamental plan. Huxley enjoyed skewering Owen's stolid defense of anatomical stability even before Darwin finally came clean with his argument for evolution in 1859.

Darwin had known all along he had no need for a salaried professional position, whether in the ranks of the clergy or in the newly emerging class of professional scientists. When he chose natural history as his lifelong passionate pursuit—probably even before leaving on the *Beagle* in 1831—his path was set. He would be a self-supported, independent scientist, following the model of such greats as Charles Lyell and James Hutton before him. The problem would then be to attract the attention and respect of the existing scientific world—a world that was now expanding beyond the narrow confines of gentleman-clergyman scientist. Darwin was trying to make it in the modern world of professional science while clinging to the lifestyle of the amateur dilettante, a lifestyle that would quickly come to be seen as incompatible with true professional science. The option simply does not exist in today's world. And that is one of the several conflicts that Darwin worked under in a world that was changing even as he lived his life.

Alfred Russel Wallace, on the other hand, did not enjoy the same set of career choices. Son of a middle-class family from Usk (west of London), Wallace was fourteen years younger than Darwin. The family had recurrent money problems, so Wallace became a tradesman, with skills in surveying (his brother William owned a surveying business), drafting, mapping, and construction. He was drawn to natural history through his love of the outdoors. After a short stint as a schoolteacher he teamed up with another young naturalist, Henry Walter Bates, with the idea of becoming a collector of natural history specimens from the Amazon region—specimens that could be sold to museums and collections to meet the demands of the increasing interest, by the 1840s, in the natural history of the remote corners of the world.

In part inspired by Darwin's account of his *Beagle* journey, Wallace and Bates were both bitten by the same bug to see the world and explore the wonders of natural history that had gotten the youthful Charles Darwin in his early twenties. Darwin had dropped a hint about evolution in the second edition of the *Voyage of the* Beagle (1845). Wallace had drawn his evolutionary ideas in part from Darwin. But Wallace, unlike Darwin, had had to fashion his own financial destiny—and Wallace and Bates paid for their own journey to Brazil, arriving in what is now Belem in April 1848.

Wallace had also read Lyell's *Principles of Geology*. Despite Lyell's lengthy arguments against Lamarck's views on evolution in the second volume of *Principles*, Wallace had absorbed the message that the earth is very old, and that processes slowly shaping its features—such as eons of erosion, slow subsidence of basins, and lifting by degrees of mountain chains—could easily have induced equally marked changes in its floral and faunal denizens. Then, too, Wallace was apparently far less skeptical and critical of Chambers's *Vestiges of the Natural History of Creation* than the majority of the professional scientists of the time. All in all, Wallace was an evolutionist—and his collecting forays were meant to take him to the places that might reveal the secrets of how life evolved.

Darwin had allowed the patterns of the natural world to seep into him, absorbing almost subliminally the key clues pointing to evolution that impressed themselves upon him during the voyage. When he left on his journey in late 1831, Darwin was as much of a creationist as any of his contemporaries; he remained so for quite a while after reading the first volume of Lyell's *Principles* in the early months at sea (Darwin received the other two volumes while he was still on board). It took a few years for him to bring the clues of evolution that had impressed themselves on his unconscious mind to the surface as explicit thought.

Wallace, by contrast, had not suffered from the same intellectual and emotional constraints. In the 1830s, he had been exposed to the thinking of the socialist utopian Robert Owen. At the very least, Wallace did not share Darwin's dread of being attacked for undermining the pillars of British society by challenging the received wisdom of the Church of England on matters concerning the development of life through time—and even seeing man as part of that evolving, living world. (Though, ironically, it was Wallace who steadfastly refused in later life—much to Darwin's consternation—to go all the way and to concede that human consciousness, the "human soul," could also have evolved through natural selection.) Wallace got his evolution more from books than from nature. Darwin got his, as we shall see, from the primal source: nature itself.

But Wallace hadn't written Darwin to say he thought that life had evolved. He was aware that Darwin himself had already gotten that far. What Wallace wanted Darwin to see, rather, was the fruits of his search for a mechanism of evolution—the intellectual quest that went along with his need to make a living that had sent him to the tropics in the first place.

Of that fateful missive, sent from Ternate, an island of what is now Indonesia, nothing survives. The original contents—undoubtedly a cover letter, plus a handwritten manuscript entitled "On the tendency of varieties to depart from the original type"—have been lost, though not before Wallace's essay had found its way into type.

Wallace's title, and the opening point of his essay, posited that, unlike domesticated animals, there is no universal tendency to "revert to type." If you look at a flock of pigeons in pretty nearly any of the world's major cities, you will see a smattering of whites, browns, and mottled mixtures of other colors. The predominant coloration is a light gray, with a darker head, red eyes, iridescent greens and purples on the neck, a darkish tail, two black stripes across the wings, and a white rump visible when the bird flies. These latter, "normal" pigeons are what the species—*Columba livia,* the "rock dove"—actually looks like in the wild. An Old World species, populations of rock doves untainted by breeding with domesticated members of their own species—i.e., truly wild doves—are hard to find nowadays (I've managed to see them only in some remote regions of the Nile Valley, and perhaps in the mountains of Spain). The oddball colors are remnants of varieties created by selective breeding by pigeon fanciers; when escaped onto the streets of New York or London, these unusually colored strains interbreed fully with all the other pigeons, and gradually the oddball colors will disappear unless more escape or are released by pigeon fanciers. Left alone, city pigeons will seem to revert back to the primal state still represented by their wild brethren.

Darwin got involved very intensively with pigeon breeding in 1855, wanting to discover in person the intricacies of selective breeding before he wrote his great work on natural selection. Wallace, in contrast, was not thinking about pigeons; but pigeons are as good an example as any of a phenomenon Wallace saw as crucial to any understanding of the nature—and origin—of species. Wallace pointed out that one of the main objections commonly raised against the idea of evolution is the sort of "reversion to type" seen in city pigeons. As Wallace himself put it in his very first sentence: "One of the strongest arguments which have been adduced to prove the original and permanent distinctness of species is, that *varieties* produced in a state of domesticity are more or less unstable, and often have a tendency, if left to themselves, to return to the normal form of the parent species." He went on to say that people assume the same is true of varieties in the wild, but then developed an argument showing that the mechanisms that produce varieties in nature make it impossible for them to "revert to type."

That mechanism is virtually identical to the idea that Darwin began developing in the late 1830s, fleshed out in greater detail and named in the unpublished *Sketch* of 1842: natural selection. The two men had come to the conclusion that species are not permanent; that new species arise from old ones in a process we now generally call "evolution" in rather different ways: Darwin through the experience of callow youth with a mind only dimly prepared by his contacts with Robert Grant and Charles Lyell's geological gradualism to absorb the lessons he learned on the *Beagle*. Wallace, in contrast, had departed England already convinced of evolution, but determined to discover its cause.

But both men had come upon natural selection in essentially the same way: they knew salient facts about variation in nature, and in animals and plants under domestication. They knew that organisms tended to resemble their parents (though neither knew why since genetics as we know it lay far in the future). And both had read the Reverend Thomas Malthus's pamphlet, *An Essay on the Principle of Population* (1798). Darwin had been exposed to Malthusian social thinking as he dined at his brother Erasmus's place when the two were living in London, after Darwin returned home from the *Beagle.* Malthus's essay was old but he was still *au courant*. Indeed, his thinking had become the justification for the "Poor Laws." Malthus thought that human population is limited by its food supply. Any aid to the poor, even during periods of famine, would end up backfiring, producing, in the end, still more people. It boggles the mind, but Malthus's thinking was especially popular among the "liberal" social reformers of the day—people like Ras's friend Harriet Martineau, who was often at dinner when Darwin was visiting his brother.

As Darwin puts it in his *Autobiography,*

> In October, 1838, that is, fifteen months after I had begun my systematic enquiry [i.e., fifteen months after he had opened his first "Transmutation Notebook"], I happened to read for amusement Malthus on *Population*, and being well prepared to appreciate the struggle for existence which everywhere goes on from long-continued observation of the habits of animals and plants, it at once struck me that under these circumstances favourable variations would tend to be preserved, and unfavourable ones to be destroyed. The result would be the formation of new species. Here then I had at last got a theory by which to work; but I was so anxious to avoid prejudice, that I determined not for some time to write even the briefest sketch of it.

Darwin says that he saw "at once" how Malthus's principle, developed originally for his analysis of the human condition, applies equally well to all species, animal and plant, in the world. Whether it was as he was literally reading Malthus's essay, or at some point soon thereafter in a moment of contemplation, he doesn't say. Though we don't know exactly when the flash of insight came, it was after he was safely back in England—either at home on Gower Street in London, or at one of the "safe" houses (of relatives or especially close friends) to which Darwin forever after his voyage more or less confined himself. Wallace, famously, in contrast put Malthus together with the natural world in a malarially induced feverish dream far from home, hardly safe surroundings.

But otherwise the intuitive, creative leap—juxtaposing some written thoughts with known details of the natural and breeding worlds—was precisely the same in both Darwin and Wallace. The details of what they knew about animals and plants in the world and under domestication were undoubtedly largely different, but they were the same in basic character. The tendency to revert to wild type—illustrated by wild and domestic rock pigeons—is the same in any domestic breed. I could have used dogs, for example, or cats for that matter. What I mean by the word "pattern" in nature refers to generalizations about phenomena, as if the very same lesson is to be drawn regardless of the details, or which species is examined. Darwin and Wallace both saw regularities in variation and inheritance in nature and in the breeders' controlled domestic venues. Both put these regular patterns together with a notion of a competitive "struggle for existence" that must occur, given the simple fact that food resources (Wallace's special favorite limiting factor) must limit the number of those organisms produced each generation within each species that could survive and sufficiently flourish, tipping the scales in their favor as the ones most likely, in turn, to leave offspring to the next succeeding generation.

Later in life, Darwin backed down from his conviction that natural selection—then as now the core of evolutionary theory—was somehow more important than the evidence that leads simply to the ineluctable fact that life has evolved. By 1858, Darwin and Wallace as well had long since thought that evolution was not in itself an original, or terribly surprising, idea, however worried and reluctant Darwin was about announcing it to the world. They both, probably correctly, thought that understanding *how* life evolves would be critical to the argument that evolution has happened; hence the supreme importance of natural selection in Darwin's mind. In fact, in his *Autobiography*, Darwin wrote that "I had always been much struck by such adaptations, and until these could be explained it seemed to me almost useless to endeavour to prove by indirect evidence that species have been modified." But in 1863, only four years after he had finally published his "Abstract"—the *Origin*—Darwin wrote in a letter to one of his earliest American supporters, the Harvard botanist Asa Gray, that "personally, of course, I care much about Natural Selection; but that seems to me utterly unimportant compared to the question of *Creation or Modification*."*

So Darwin did see the difference. And if it can be argued (as it often is) with some

*Here, and elsewhere passim, the italics are in the original text unless otherwise identified.

plausibility that he managed to convince the world of the elemental fact of evolution at least in part because he had a mechanism (natural selection) for it, the philosopher David Hull is also right to point out that, at least in the scientific world, the notion of evolution was nearly universally accepted immediately upon publication of the *Origin*, though natural selection remained a bone of serious scientific contention. It was the patterns in nature, the sequence of fossils in time, and the patterns of geographic variation in living species of animals and plants (his "indirect evidence") that in the end did the trick in convincing the world that life had evolved, even as they had come nearly a quarter century earlier to convince Darwin of the fact of evolution.

The story of Darwin's creative scientific genius lies in how he derived the very fact of evolution, how he became convinced that species are not immutable but rather give rise to one another, well before his discovery of natural selection.* Between 1838 and 1842–44, the two components—evolution as fact and natural selection—had a curious, highly illuminating interactive history in his mind, giving us rare insight into scientific creativity, and even the relation of the subconscious and analytic parts of the scientific thinking process (or any other, for that matter). Darwin's notebooks, unpublished manuscripts, letters, and ultimately his publications (chiefly, of course the *Origin* but also *The Descent of Man*) come very close to revealing creativity in action.

What Darwin meant by "a theory by which to work" was no less than taking natural selection and trying to derive—as "predictions"—the expected consequences of natural selection in action over long periods of time. From natural selection, Darwin tried to derive those very same basic patterns that he had seen in the natural world, including the original three that had led him to evolution in the first place. But first he had to have stumbled on those patterns and have them emerge as conscious realizations.

When Darwin set out to sea on the *Beagle*, he was still very much of a creationist, comfortable with the arguments of the Reverend William Paley, a part-time lecturer at Cambridge, that the Almighty was responsible for the appearance of new species. In a metaphor still very much alive in the writings of the architects of modern-day creationism's "Intelligent Design," Paley had written in his *Natural Theology* (1802) that chancing upon a watch lying on the ground would automatically imply to the finder not only that someone owned the watch and had accidentally dropped it, but that the

*The psychologist Howard Gruber has also analyzed Darwin's notebooks and manuscripts from the point of view of Darwin's creativity (Gruber, 1974).

watch surely had a maker. The intricacies of workmanship and construction were such that the watch must have been conceived and built by an intelligent designer—in this case, a human watchmaker. So too, Paley argued, with Nature's productions: the complexities of an animal's body can only be explained by its having been conceived and fashioned, not just by its progenitors, but by the Creator Himself. Only God can make a tree. While still a young man at Cambridge, Darwin pronounced himself charmed, delighted, and utterly convinced by Paley's argument.

But, of course, he was also ambitious—determined to "take a fair place" among the newly emerging ranks of the men of science. He was conscious of his geological ambitions, and had brought to the voyage the trick of turning unfamiliar territory into an open book where the geological history of the region could be read and analyzed through observation and deduction.

That was the extent of Darwin's prepared mind. During the voyage he spent as much time as he could on land—to see the culture, but especially to study geological strata and to collect the rocks, minerals, fossils, and, increasingly, the flora and fauna of South America. And, of course, to escape the perpetual bouts of seasickness aboard cramped quarters.

Here we have an impressionable young man (Darwin was only twenty-two years old when the *Beagle* set sail), keen on natural history and, despite his devotion to Paley's *Natural Theology*, eager to learn all manner of things. But there was something else about Darwin that set him apart from most other scientists—then and now. He always thought that, fascinating as the details about the breeding behavior of primroses, the odd anatomies of gigantic South American fossil mammals, or the pattern of variation of mockingbirds within the Galapagos Islands might be, it was what these "facts" told him about the nature of the world in general that was the true intellectual prize of such often laboriously painstaking investigations. He loved them both: the meticulous collection of facts and (later in life, at Down) the direct experimentation; but he especially rejoiced in what these facts told him *in general* about the natural world.

Darwin saw this in himself—at least as he grew older and could reflect on what he actually had accomplished. In his *Autobiography*, he writes that he saw himself as working "on true Baconian principles, and without any theory collected facts on a wholesale scale," such that "my mind seems to have become a kind of machine for grinding general laws out of large collections of facts." When Darwin entered the scene, Baconian induction was still the model of how science was to be done: one has no prior idea, no prejudice about what to expect but rather goes to the natural world unfettered with presuppositions of the way things are and follows the rule laid out in Ecclesiastes: "Speak to the Earth and it shall teach thee."

Already in Darwin's time, Baconian induction was beginning to lose its grip, especially on the new class of young paid professional scientists. Most telling of all is probably yet another thought from Darwin's *Autobiography*: "I have steadily endeavoured to keep my mind free so as to give up any hypothesis, however much beloved (and I cannot resist forming one on every subject), as soon as facts are shown to be opposed to it." Darwin, like any good modern scientist, wanted to be the first among his peers to know that his idea is wrong: It is much better to find out for yourself than to have someone else point it out.

Scientists today are in total agreement with the biologist Peter Medawar, who wrote in 1969 that "innocent, unbiased observation is a myth." Rather than those old Baconian principles, now seen as hopelessly naive, scientists claim the "hypothetico-deductive" method as the core of the scientific process. One does have, in fact, a hypothesis firmly in mind (or perhaps two or even more alternatives). One then proceeds to make predictions about the nature of the world—what *ought* to be observed if that hypothesis is true. And, in one of the nastier turns of logic in human affairs, it is also universally agreed that finding those predictions to be true does not ipso facto make the hypothesis itself true. If the predicted observations do not turn up—if in fact something else is observed—then the hypothesis is false (unless there was an error in deriving the predictions from the hypothesis, which in my view is what happened to Darwin in one area of his thinking). One can falsify a hypothesis but one can never prove it. The best that can be done is to make more and more predictions, and find them confirmed over and over again. One then sees that the proposition has been so highly confirmed that it has, as Darwin himself would have said, itself become a fact—though in principle it is still falsifiable.

Consider the proposition that the Earth is round (it is, "in fact," an oblate spheroid). The Greeks thought so, if for no other reason than the top of the mast is the first part of a ship to appear to keen eyes as the ship sails toward port. But to many people, the idea that the Earth is ball-shaped was still difficult to accept when Columbus set sail westward to find an alternate route to the Indies. That Columbus and many who followed in explorations around the world failed to fall off the edge quickly led to confidence that the world is indeed round. And despite the persistence of some flat-earthers (who insist that Moon shots were faked), there are many pictures from the upper atmosphere and from the Moon itself amply confirming, for all to see, that the Earth is indeed a sphere.

So there is a spectrum of hypotheses: from rank speculations, many of them fanciful and easily falsified, to good, educated guesses—ideas that have stood repeated testing and have consistently been corroborated. Their predictions have never been falsified. The roundness of the Earth is one such idea. So is evolution—and it was Darwin who showed the way.

Michael Ghiselin, in *The Triumph of the Darwinian Method* (1969), takes a long, hard (though unabashedly admiring) look at Darwin's methods embedded in his scientific work. Ghiselin, a biologist, makes a strong case that Darwin was indeed among the first to make conscious use of what is now called the hypothetico-deductive method, and he agrees with other historians and biologists that, despite what Darwin himself said about his methods, he was no naive inductivist, no slavish follower of Baconian principles. Indeed, Medawar is undoubtedly right: it is impossible for the human mind to take on observations without some form of ordering of that information, and that ordering is likely at least in part to come from thought pathways already there.

But it is a mistake to trumpet Darwin's achievements solely in the light of his modern hypothetico-deductive bent. For all the vaunted worth of this method in science, it pays utterly no heed to where hypotheses come from in the first place. It is all well and good to test hypotheses already in place—in one's own mind, or, more commonly, put forward by others. The hypothetico-deductive method—the recipe for the careful observation and experimentation that is indeed a requisite of all good science—is in fact rather bland and soulless. It is the stuff of humdrum science, important as it may be. It is not where the grand ideas come from. Indeed, though some scientists see this method as the essence of science, effectively setting science apart from all other spheres of human thought, in reality it is simply common sense rigorously applied to the material world. One buys a car the same way: one tests the "hypothesis" that a certain car would be a good buy by test-driving it; by comparing its features with what one thinks one needs in a car; and, if it is used, by kicking the tires.

What is really interesting is where ideas come from in the first place—especially the grand ideas, not only in science but in culture generally.

Darwin, perhaps more than anyone else I've met or read, knew his own mind, and there is every evidence that, naive or not (and, biologically speaking, he was naive when he left England on the *Beagle*), he pretty much knew, at least in retrospect, how he stumbled onto evolution in the first place. It was by letting patterns seep into his mind, virtually unknowingly, only to have them surface and, in due time, become conscious thoughts pointing to a grand generalization or "law": evolution. When he consciously saw what these patterns were, he looked for others. And then, when he had become convinced that life had evolved, he turned the tables and asked: What would we expect to see if life had evolved? And especially after he had thought out natural selection by the late 1830s, he went even further and tried to derive his patterns—his "indirect evidence"—of evolution from natural selection itself.

Darwin embodies what I think goes on in all great, creative scientific minds: patterns impinge upon the brain which suggest both a question and an answer simultaneously. The answer, once grasped, makes it possible to see the initial patterns in nature as "expectations" (as Darwin himself was fond of calling them). What were the patterns that impinged on his brain while on the *Beagle* (begging the question, for the moment, when and where he was at the moment evolution actually crystallized as a coherent, conscious idea)? Darwin himself tells us in four critical passages. The first was pointed out by Francis Darwin in his Introduction to *The Foundations of the Origin of Species*, his invaluable, annotated edition of his father's 1842 *Sketch* and 1844 *Essay*, published in 1909 to celebrate the centennial of his father's birth:

> . . . in his Pocket Book under the date 1837 he [i.e., Charles Darwin] wrote "In July opened first note-book on 'transmutation of species.' Had been greatly struck from about month of previous March on character of South American fossils, and species on Galapagos Archipelago. The facts origin (especially latter) of all my views."

The second passage comes right before the sentence about "confessing a murder" in Darwin's letter to Joseph Hooker of January 11, 1844:

> Besides a general interest about the southern lands, I have been now ever since my return engaged in a very presumptuous work, and I know no one individual who would not say a very foolish one. I was so struck with the distribution of the Galapagos organisms &c.&c., and with the character of the

American fossil mammifers &c.&c., that I determined to collect blindly every sort of fact, which could bear on any way on what are species. I have read heaps of agricultural and horticultural books, and have never ceased collecting facts. At last gleams of light have come, and I am almost convinced (quite contrary to the opinion I started with) that species are not (it is like confessing a murder) immutable.

The third passage, appropriately enough, opens *On the Origin of Species*:

> When on board H.M.S. "Beagle," as naturalist, I was much struck with certain facts in the distribution of the inhabitants of South America, and in the geological relations of the present to the past inhabitants of that continent. These facts seemed to me to throw some light on the origin of species—that mystery of mysteries, as it has been called by one of our greatest philosophers.

And finally, in his *Autobiography* (p. 52), Darwin is more specific, breaking the patterns down into three sets:

> During the voyage of the *Beagle* I had been deeply impressed by discovering in the Pampean formation great fossil animals covered with armour like that of the existing armadillos; secondly, by the manner in which closely allied animals replace one another in proceeding southwards over the Continent; and thirdly, by the South American character of most of the productions of the Galapagos archipelago, and more especially by the manner in which they differ slightly on each island of the group; none of the islands appearing to be very old geologically.

That Darwin was not overtly looking for these patterns is nicely borne out in his own accounts of finding them accidentally—while literally ingesting the very specimens that embody two of these three patterns. While Darwin indeed ate armadillos, thereby giving him the opportunity of intimately comparing their anatomy with that of *Glyptodon*, a gigantic shelled member of the primitive mammalian order Edentata (as are armadillos—hence Darwin's use of the term "allied," to a certain extent begging the evolutionary question), he was no doubt deliberate in his comparisons. He was struck, as he himself says, by the fact that a gigantic extinct form, though in no way a direct ancestor to armadillos, but nonetheless clearly of the same general mold as armadillos, existed in the Tertiary rocks of southern Argentina. Why would a Creator produce modern, similar forms of some extinct species but confine their appearance solely to South America? (Armadillos do get into North America as well, but Darwin did not know that.)

But he literally was eating the evidence of his second pattern when he suddenly saw—almost too late—what was happening. Darwin had heard from some of the Argentinean ranchers and gauchos he had met of the existence of a second rhea (what he refers to as the South American "ostrich"), said to be smaller and somewhat differently colored and behaved from the common rhea of the Pampas. One day, while eating rhea for dinner on board the *Beagle*, Darwin and his messmates had almost consumed what he thought was just a small, immature common rhea when he realized with a start that he was in fact dining on the smaller rhea—which he had been looking for but had not as yet found. He managed to save the head, neck, a wing, the legs, and some larger feathers, and shipped the remains back to England, ultimately to be named as a distinct new species, *Rhea darwinii*. (Many of the species named after Darwin were based on specimens he collected while on the *Beagle*, but as luck would have it, specimens of *Rhea darwinii* had already been collected and named by the French explorer, collector, and naturalist Alcide d'Orbigny. As further evidence of Darwin's competitive nature, he had been distressed to learn that d'Orbigny had passed through many of the same locales that he himself visited, fearing that d'Orbigny would discover all the important animals, plants, and fossils before him. In the case of "Darwin's rhea," his paranoia turns out to have been amply justified.)

There is no reason to think Darwin made anything out of the remains of the smaller rhea as he went about scavenging its remains from the dinner table other than confirmation of a different species—a different sort of rhea, a species he had heard rumors of, now confirmed. When he saw this as a generalized pattern of large-scale replacement of "closely allied" species (the two rheas overlap slightly in their distributions in South America), it is impossible to say. But again, the question is: Why would God choose to replace one perfectly good species with another closely similar one, as one traverses the length of the South American continent?

The dining table figures once again into Darwin's third general set of patterns leading to his inevitable conclusion that species cannot be immutable. Apparently the gov-

Darwin thought that the Galapagos tortoises were placed on the islands by sailors as a future source of food. He ate them, but only later began to take seriously what the Spanish governor had told him: that it was possible at a glance to identify which island they came from. He was delighted when he later found out that the tortoises are undoubtedly native to the Galapagos.

ernor of the Galapagos Islands told Darwin late in his visit that there were separate varieties—or separate species—of tortoises on the various different islands. He and the crew had been capturing and eating tortoises for dinner (he brought one small one back to England) in the time-honored maritime tradition. (The Great Auk of the northern Atlantic is said to have been driven to total extinction by the 1850s—depleted by countless sailors marching them up the gangplank and into their larders.) The distinctively different-by-island carapaces were unceremoniously dumped overboard as garbage. Darwin assumed that prevailing opinion was correct: that sailors had imported these tortoises from elsewhere in the Pacific, stocking the islands for future food resources. Apparently it was only after leaving the Galapagos that he took the pattern of inter-island tortoise variation seriously.

Darwin also, famously, missed the patterns of variation in the thirteen species of

Darwin's rhea, these days more properly called the "lesser rhea." This is the species Darwin had heard about; he had to rescue the remains of the only one he ever found from dinner scraps.

The small ground finch, one of the thirteen Geospizine, or "Darwin's," finches in the Galapagos Islands. When he visited the Galapagos, Darwin did not realize that the small black and greenish birds form a group of "allied" finch species, and failed to keep accurate notes and labels for the finches he did collect. It was not until after he reached home that the ornithologist John Gould realized the thirteen species of Galapagos finches are closely related.

what are now collectively known as "Darwin's finches" on the Galapagos Islands. Indeed, the birds are generally drab and difficult to identify, though Darwin, in inspecting their beaks predominantly, thought them to belong variously to a number of different groups (in other words, not a coherent assemblage allied to finches). He failed to keep notes on the islands from which his collections of these birds came. Only after John Gould looked at all available specimens and told Darwin that the birds all belonged to the same group did Darwin realize that, like the tortoises, he had missed a perfect opportunity to document variation within and between the different Galapagos Islands.

But Darwin did *not* miss the mockingbirds. He saw that different islands had different types of mockingbirds—though whether they were all varieties of one species, or distinct species, he couldn't tell. (Indeed, as we shall see, the difficulty in telling "varieties" from "true species" when comparing similar forms from separate, neighboring locales is, as Darwin would say, exactly what you would expect to find if evolution is taking place.)

This is what Darwin had to say about those Galapagos mockingbirds—in a separate notebook entitled "Ornithological notes," written long after the *Beagle* had sailed away from the Galapagos. These are his earliest words yet found that hint of his tumbling to the very idea of evolution; though occurring in the section specifically on mockingbirds (which in his notes he refers to by the common name Thenca), he also remarks on the tortoises:

> These birds are closely allied to the *Thenca* of Chile. . . . I have specimens from four of the larger islands. . . . The specimens from Chatham and Albemarle Isd appear to be the same; but the other two are different. In each Isld. each kind is *exclusively* found; habits of all are indistinguishable. When I recollect, the fact that the form of the body, shape of scales & general size, the Spaniards can at once pronounce, from which Island any Tortoise may be brought. When I see these Islands in sight of each other, & possessed of but a scanty stock of animals, tenanted by these birds, but slightly differing in structure and filling the same place in Nature, I must suspect they are only varieties. The only fact of a similar kind of which I am aware, is the constant/asserted difference—between the wolf-like Fox of East and West Falkland Islds.—If there is the slightest foundation for these remarks to zoology of Archipelagoes—will be well worth examining; for such facts would undermine the stability of Species. (Galapagos MS 73)

Nora Barlow, who published these notes in 1963, believes that they were written in September or October 1835—nearly a full year before the *Beagle* returned. Current thinking sees these notes as written in the summer of 1836 as the *Beagle* headed for home. In any case, it is clear that Darwin was already contemplating the idea of evolution while still aboard the *Beagle*. He was well on his way toward full realization about what these seemingly casual observations might really mean—aside from the novelty, thrill, and their own intrinsic importance to science and therefore to his nascent career.

Why would God want to put different mockingbirds, different tortoises (and, he was much later to find, very different plants) on the different islands of the Galapagos—and other archipelagos?

That's the trick of creativity: to get the observations organized sufficiently at the subconscious level, and then to be able to bring them out, sort them out, write them down *consciously*. That's the creative side. Once he explicitly, consciously realized the nature of these three sorts of patterns—the fossils, large-scale geographic patterns, and smaller-scale patterns of variation from island to island—in each situation, with closely similar, albeit distinct, forms replacing one another, Darwin had to ask why? Why would God want to do that? Wasn't there a simpler, natural explanation?

Darwin first intuited, then consciously realized, each of these three patterns. He always described them as general: "the manner in which closely allied animals replace one another in proceeding southwards over the Continent," for example, was initially based (so far, at least, as can be told by his notes and, later, Transmutation Notebooks) *on a single example*: the two rheas. When he got home he started to read, and quickly compiled many more similar examples from all over the world, based on the work of others. *But he stumbled onto the pattern of geographic replacement from one example!*

As far as fossils were concerned, he had a few other examples: the extinct giant ground sloth is a forerunner (albeit not a direct ancestor) of the two existing sloth species still extant in South America. He also thought (as it turns out, erroneously) that the large skull of the mammal *Toxodon* represented an extinct relative of the South American capybara, the world's largest living rodent. Richard Owen misled Darwin with his erroneous conclusion that the fossil *Macrauchenia* was a related predecessor of the South American camel (llama). That's it: four examples, only two of which turned out to be valid.

And as far as the Galapagos are concerned, he had the mockingbirds—and maybe the tortoises. Not the finches, and the plant evidence was only to come long after he reached home.

But these few examples were enough to hint at the existence of regular, repeated patterns—through the entire geological history of life and over the entire surface of the globe. He went on to compile so many examples that he hardly mentioned his primal cases in his polished writings, especially in the *Origin* itself. But he did acknowledge them: great intuitive leaps made by a young man, hardly more than a boy, who indeed was far more a Baconian inductivist, very impressionable, when Nature literally cast her productions on his mind—than the meticulous hypothetico-deductivist he would later blossom into. As a callow youth (as my colleague Joel Cracraft has remarked), Darwin "let Nature come to him." It was only after he arrived at the truth of evolution through intuition and induction that Darwin turned to rigorous hypothesis testing—starting with his theory of evolution through natural selection.

Take a walk, if you can possibly get to Down House someday, along the quiet reaches of the Sandwalk. You can almost feel Darwin walking along, thinking of cowslips, or seeds he'd sent for in the mail that had not yet arrived. Or the reason why so many different species of plants occupy the neighboring field. (A survey he did perform, with the help of one of the servants, which represents one of the earliest biodiversity surveys of its kind on record.) You can feel that even as he grew older, he didn't stop being an intuitive person, someone who was really good at having thoughts impinge on what had long since become a prepared mind—novel thoughts that he had the unusual ability to bring to the surface and take a cold, hard, analytical look at. Still, they started as intuitions.

The difference between the young fledgling naturalist and the grizzled old famous scientist was one of degree, not kind. On the *Beagle*, wonderfully, Darwin was intuitive and open as he painstakingly collected his facts; as he grew older, he became more adept at reasoning, predicting, experimenting—trying above all else to be very sure about the facts. But he never lost sight of what, in the end, it had always been all about: finding some very basic "laws" about the way nature characteristically behaves, over and over and over again.

The result: evolution, the idea that all life on earth, from time immemorial, is descended from a single common ancestor. A result so powerful that it ranks among the great ideas in the history of the Western world.

And you can feel, as you walk along the Sandwalk, Darwin's anxiety—the roots of his fear that he would be reviled for undermining religion, a central pillar of British life. For Darwin's notebooks make it clear that the only rival hypothesis he could think of—the only one that was there to rival the notion of organic evolution through natural causes—was in fact the Creation story of revealed and received Christian religious doctrine. Time and again he asked rhetorically why a Creator would make so many beetles; arrange different plants and animals on nearby islands; replace fossil forms with closely similar living forms found nowhere else in the world. Darwin had quietly been taking on religion—but not telling anyone, at least beyond a very small circle of friends, mostly scientists.

And now here was Alfred Russel Wallace, evidently with no such compunction, about to steal his thunder. Darwin made up his mind, as he ambled slowly back to his house, to do the right thing, and send Wallace's manuscript on to his friend Charles Lyell. Soon thereafter he wrote to Lyell.

Down, 18th [June 1858]

My dear Lyell,

Some year or so ago you recommended me to read a paper by Wallace in the *Annals*, which had interested you, and, as I was writing to him, I knew this would please him much, so I told him. He has to-day sent me the enclosed,* and asked me to forward it to you. It seems to me well worth reading. Your words have come true with a vengeance—that I should be forestalled. You said this, when I explained to you here very briefly my views of Natural Selection depending on the struggle for existence. I never saw a more striking coincidence; if Wallace had my MS. sketch written out in 1842, he could not have made a better short abstract! Even his terms now stand as heads of my chapters. Please return me the MS., which he does not say he wishes me to publish, but I shall of course, at once write and offer to send to any journal. So all my originality, whatever it may amount to, will be smashed, though my book, if it will ever have any value, will not be deteriorated; as all the labour consists in application of the theory. I hope you will approve of Wallace's sketch, that I may tell him what you say.

My dear Lyell,
yours most truly,
C. Darwin

Darwin's friends, Charles Lyell and Joseph Hooker, intervened. Together they proposed sending not just Wallace's MS but also extracts from Darwin's 1844 *Essay*, together with an excerpt from a recent letter (October 1857) that Darwin had written to Asa Gray, to the Linnaean Society. Hooker and Lyell inserted the Wallace/Darwin presentation into the agenda, and the paper was read out to the meeting on July 1, 1858. Darwin, unsurprisingly, stayed home, while Wallace, of course, was still far away.

*NB: Historians disagree on when Darwin actually received the letter from Wallace.

The joint production ensured that Darwin's priority was thus gently, if unmistakably, established. For his part, Wallace on the whole seemed rather pleased by the turn of events; after all, such rapid, albeit joint, publication (on August 20, 1858) in a prestigious scientific journal was more than he could otherwise have hoped for.

In summing up that year's events, the president of the Linnaean Society remarked that not much new had occurred. He was wrong.

CHAPTER 3

Darwin's Evolution: Issues, Contexts, and the Red and Transmutation Notebooks

What is "evolution"? Though Darwin used the term "evolved" in one or two places in the *Origin*, only later in his life did the actual word "evolution" come into vogue—to stand for "transmutation," or even Darwin's more mellifluous if more cumbersome "descent with modification." Paring it down to its barest bones, evolution is the fate of transmissible information over time. Darwin may have known nothing about the nature of what we now call genetics—the chemistry and structure of DNA, RNA, the various parts of chromosomes, the notion of "genes"—but he did know that organisms resemble their parents; that the variation in the appearance of organisms within a single species is heritable; and that more organisms are produced each generation than can possibly all survive and themselves reproduce. That was enough for him (and, of course, for Wallace) to derive the theory of natural selection. Natural selection is one biological mechanism that regulates the transmission of whatever forms of information make it through to the next generation.

Defining evolution as the fate of transmissible information invites comparison with other, non-biological systems. For example, humans learn from each other by transmitting the rules of culture to one another and across generations in ways that are distinctly "evolutionary"—and with implications for resolving the Reverend William Paley's edict that anything with apparent complexity of design in nature must necessarily presuppose the existence of a supernatural Designer.

We'll return to the problems of design evolution in the final chapter, as they have direct relevance to the still-festering creationism controversies revolving around issues of complexity and (apparent) "Intelligent Design." The questions raised by the simple definition of evolution as the "fate of transmissible information within a system," specifically biological systems, are crucial, as they provide the conceptual hat rack to understand how Darwin got his ideas in the first place; what he did with them (and they changed a lot!) in the twenty-odd years before he summoned the courage or was prodded into publishing them in the *Origin*; and what we are to make of the rather labyrinthine history of evolutionary thinking since Darwin's time—including especially understanding the initial impact of each of the two "genetic revolutions" that have occurred since Darwin's day. Thinking about evolution as the stability and change of information through time and over space enables us to develop a guide to understanding most of the debates in evolutionary thinking—from pre-Darwinian times up through today.

Information in Evolution. What do we mean by "information"? It is the instructions for the production of component parts and how they are put together—whether we are talking about the human body or a human-made artifact like a computer. Naturally, this raises the questions: what carries this information, how is it transmitted, reassembled, and turned into new objects, whether human babies or more computers? Darwin thought that the answer to his search for a mechanism of evolution—a search begun in earnest not long after he arrived home from the *Beagle* in late 1836—would come from the still largely unknown processes of heredity. Evolution, in other words, might reasonably be thought of as a change in heritable information. And, as we shall see in a bit more detail below, though Darwin's ideas about how the hereditary process works have long since been shown to be dead wrong, enough was known about patterns of heredity that he could still learn from experiments that the flow of information can be controlled to some degree by breeders.

In fact, as Theodosius Dobzhansky first pointed out in the 1930s, the details of how the genetic process works are sublimely irrelevant to understanding how natural selection works. Just as I do not fully understand in detail how my touching the letter *S* on the keyboard of my computer is actually translated into electrical impulses that show up as code interpreted by my word processing system as the letter "S" on my screen, I don't really care. It is just important that it is so, and that it works for all the other electronic and printed versions of my writings. Dobzhansky made a similar point about natural selection: what is needed is (1) heritable variation in (2) a population whose size is limited by some extraneous factors (like food availability), such that (3) there will be a bias in the transmission of that information no matter what the precise details of how organisms come to resemble their parents. As Dobzhansky said, natural selection is a process of differential survival of genetic information among individuals within a population—and has nothing directly to do with the details of how genetic information is transmitted from parent to offspring in the first place.

That is not to say that the details of modern genetic understanding of the genome, how it is translated into a developing organism, how that information is replicated and passed along to the next succeeding generation—and how the components of that information are modified or remain unchanged—are irrelevant to studying evolution. Some of the most challenging episodes in the history of science have come about when new fundamental understandings of the hereditary process have emerged, as they did in a brief explosion of work in the first decade or so of the twentieth century triggered by the discovery of Mendel's Laws; and then again later in midcentury with the resolution of the structure of the DNA molecule and, shortly thereafter, the cracking of the DNA code. The problem in both those instances was to see how this new understanding fit in with existing ideas about the evolutionary process—intellectually rough and challenging episodes in both cases that generated a great deal of controversy.

But Dobzhansky's basic insight—that the genetics of populations is not the same as, nor wholly dependent upon, the specific genetic processes that go on within individuals—raises an important point. Information (we now call it "genetic" information) comes packaged at different levels of biological organization. The genome—all the genes on all the chromosomes, including the vast amount of structural and non-coding elements present inside each cell—is a complex miniworld in which some elements control when others are turned off and on, and in what sequence. Mutations (in many cases, simple copying errors when chromosomes replicate as cells divide) mean that there can be variation in genetic information within the cells (billions of them in complex organisms) of a single individual. And, identical twins notwithstanding, for the most part no two individuals have exactly the same genetic information.

Continuity vs. Discontinuity of Information: The Role of Geography. This Darwin and his contemporaries knew: organisms vary inside populations, or in the barnyard, where breeders would select and allow to breed only those individuals with traits they wanted to see further enhanced. But that wasn't all: populations of what otherwise seemed to be the "same" species appear to differ in certain traits from other populations living elsewhere. There might be a prevalence of darker coats in squirrels from some cities, as is the case for eastern gray squirrels in North America. There are far more black squirrels in Toronto than in most populations living in New York; yet dark and lighter gray squirrels interbreed freely, and dark and lighter individuals appear in the very same litter, implying that their differences are slight hereditary variations within the same species.

Consider the birds in the nearest vacant lot or backyard. The house sparrows, starlings (both European invaders in North America), chickadees, finches, and crows are all readily distinguishable one from another. They tend to look pretty much alike, each of the different members of the "same species," though you might notice small differences in size and plumage color in some of them. But crows are crows, starlings are starlings, and so on. And if you took the trouble to check them out in springtime, you would find that mating was going on strictly between members of the opposite sex within each of these distinct groups: crows make baby crows, starlings baby starlings. You get the distinct impression that there are different "kinds" of birds—and mammals, and insects, plants, and so on—living cheek by jowl, whose individual "members" all look pretty much alike, but are separated from other kinds of animals or plants by what some naturalists of the past have referred to as "bridgeless gaps."

As you turn your bird-watching gaze further afield, though, things don't always seem as neatly clear-cut as they did back home. The chickadees all look pretty much alike across North America—at least in the more northerly states—though you begin to notice differences in their calls and songs. The ones in the southeastern states seem a bit smaller, their voices more shrill and their song a bit more rushed. Worse, the chickadees in San Francisco have brownish backs, the ones in the Utah mountains have a black stripe through the eye, and the ones in Canada have brown, not black, on the top of their heads. Chickadees vary geographically. The question then becomes: how do we tell if the chickadees in Ohio, or in South Carolina, or Utah, San Francisco, or Canada, belong to the "same" group of chickadees seen in your backyard? What are the limits of geographic variation? Why do different forms—sometimes starkly different, at other times more gradationally so—of what otherwise appears to be the "same thing" appear to *replace* one another in adjacent regions? Is it all arbitrary, what we call a "species"? What are the limits to a species: when do we say something is a variety of the same species living elsewhere and when do we say they are different species?

What are "species," anyway? Darwin saw the original problem of evolution as coming up with a solution to Herschel's "mystery of mysteries"—the origin of new species. It is no surprise, then, that the key to understanding much of the history of evolutionary thinking revolves around what people have thought species are. Unsurprisingly, all manner of answers have been given, ranging from one extreme which insisted that species are strictly fixed entities, to the other—that species don't exist at all, but rather are strictly human concepts, artifacts of the psychological urge to pigeonhole nature into discrete entities that do not, in fact, exist. The heart of the question has always been: are species natural entities?

So discreteness versus continuity is a major issue in evolutionary circles—from the pre-Darwinian resistance to evolution in the emerging ranks of naturalists and early professional biological scientists, right on up to the present day. And if species have been the focus of much of the attention in terms of such issues, it is also important to recall that similar debates have occurred at different scales: is inheritance discrete or blending, for instance? Is variation within populations continuous or discrete? For example, Dobzhansky, when he made the point in his epochal *Genetics and the Origin of Species* (1937) that the rules of genetics of individuals ("physiological genetics") were different from the rules governing the genetics of populations ("population genetics"), concluded that mutations are discrete, either/or phenomena. Natural selection, on the other hand, modifies the frequencies of different genic forms ("alleles") in populations, so variation appears to be smoothly continuous within populations. And though perhaps we are getting ahead of the story a bit, it is the case that Dobzhansky wrote his book to stress that discontinuity appears once again at the next highest level: species.

Continuity vs. Discontinuity in Geological Time. This is the key question: continuity versus discontinuity when it comes to heritable differences between individuals, between populations, between "species," and ultimately between larger-scale groupings of organisms, especially when we compare variation from place to place. Now consider the role of time. The question of discreteness versus continuity also arises in issues posed by newly appearing "species" replacing older ones gone extinct: Herschel's—hence Darwin's—"mystery of mysteries." The fossil record has hauntingly similar analogues to the discreteness/continuity and replacement patterns we see geographically in the modern world. Only more so, because the fossil record also lets us see replacements and issues of discreteness versus continuity in space as well as in time. Under favorable conditions, it is possible to study the evolutionary history of a single lineage showing patterns of within-population variation, as well as variation over space at the same time, and variation, as well, through time (now often measured in the millions of years—time scales undreamt of when Darwin set sail on the *Beagle* in 1831).

But geological time introduces yet another set of issues: rates of evolution. How fast can and does evolution happen? Mixed in with this relatively straightforward question of evolutionary tempo are trickier questions: does evolution proceed at more or less constant rates over long periods of time? And what, exactly, do we mean by "long periods of time," anyway? The late paleontologist Stephen Jay Gould referred to this as a problem of scale: what seems like a short period of time to a geologist may seem tremendously long in terms of a human's lifespan—and thus to the eyes of a biologist accustomed to measuring rates of evolution over the course of a few years.

Rates of evolution are easily confused with issues of continuity versus discreteness. To choose an example that I have grappled with my entire career, consider the rapid rates of evolution that underlie the appearance of new species in the fossil record—part of the pattern that Stephen Jay Gould and I called "punctuated equilibria." Gould and I believed that the long periods of time (typically hundreds of thousands, even millions, of years), with little or no measurable change shown by most species known from the fossil record, were "punctuated" by rapid episodes of evolution that gave the appearance of abrupt discontinuity between older species and new ones that replaced them (shades of the "mystery of mysteries"). We were accused of being anti-Darwinian, advocating a form of abrupt, sudden change (often called "saltationism," from *saltus*, the Latin word for "jump") instead of a smoothly gradational, continuous form of change between ancestral and descendant species. It truly hurt to be called anti-Darwinian!

But we were not anti-Darwinian heretics. We always said the transition was "rapid" compared with the vast periods of evolutionary non-change (we called it "stasis"). But it depends on what is meant by "rapid": we made every effort to be clear, and kept saying that our best estimates of "rapid" were in the "five- to fifty-thousand-year range"—rapid by geological standards, but within the range of known evolutionary rates back in the

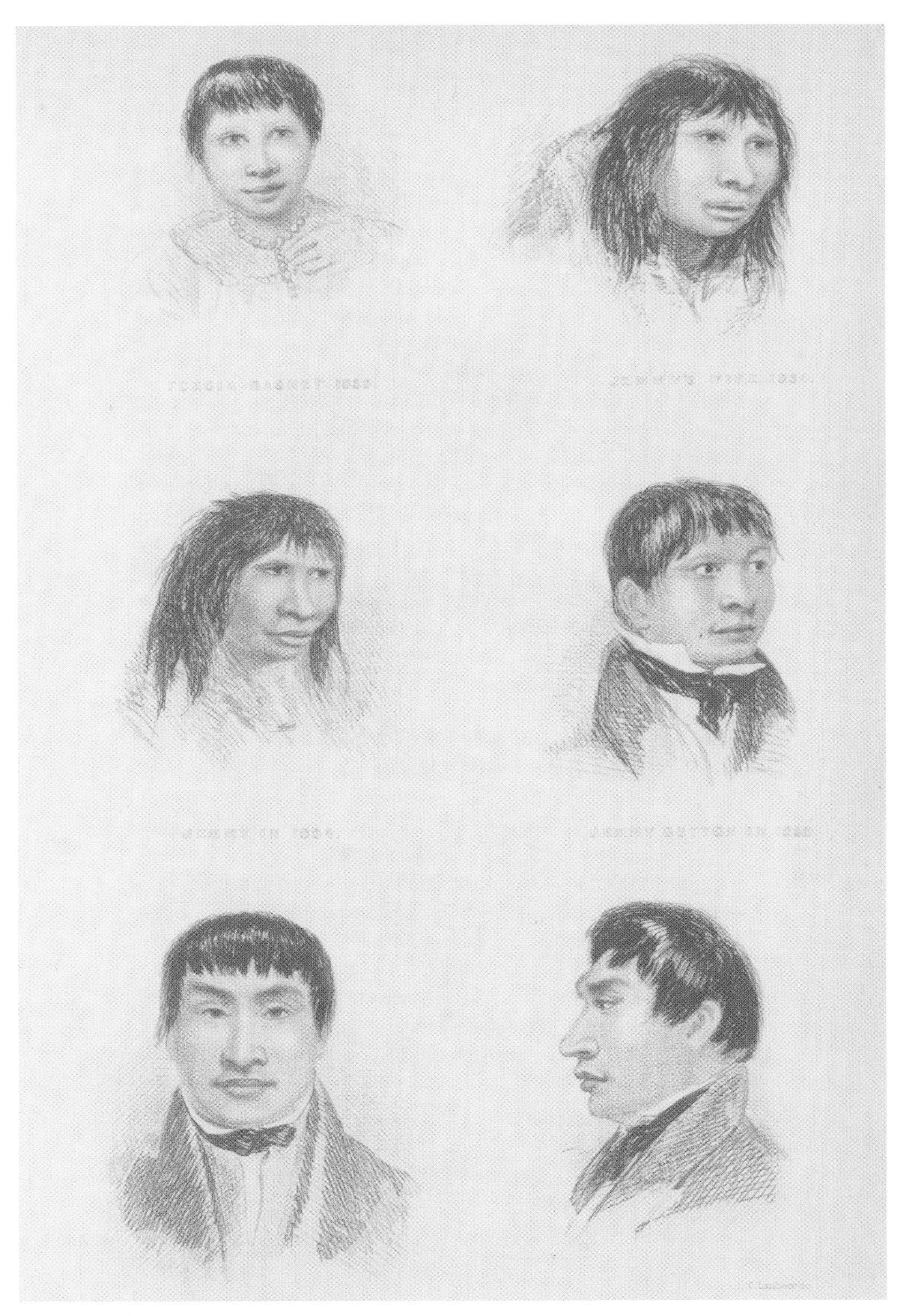

The Fuegians who traveled home on the *Beagle*: Fuegia Basket (*top*), Jemmy Button (*middle*), and York Minster (*bottom*). Sketch by Captain Robert FitzRoy.

A basalt flow (cooled igneous rock) and associated sedimentary rock layers on the Santa Cruz River where Darwin's fossil finds fueled his evolutionary insights.

A Patagonian tableau—with rheas, cavys, and other elements of the local fauna that led Darwin to his evolutionary ideas.

The *Beagle*, with accompanying ship *Adventure*, at Mount Sarmiento, Magdalen Channel, Straits of Magellan. Conrad Martens.

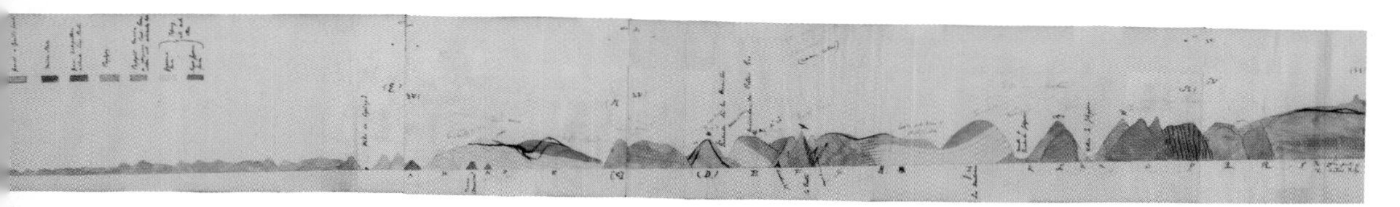

Darwin's geological cross section of the Andes. His paper on the elevation of the seafloor to form the Andes was his first scientific contribution after his return home, enhancing his growing reputation as a geologist of merit, and furthering his ambition to "take a fair place" among the men of science in London.

The ruined cathedral of Concepción after the earthquake that had thrown Darwin to the ground while walking in woods further to the south. Earthquakes, volcanoes, and tidal waves are rare jolts that punctuate earth history, providing uneasy counterpoint to the gradualism latent in Lyell's doctrine of uniformitarianism.

Marry | This is the Question | Not Marry

Children—(if it Please God)—Constant companion, (& friend in old age) who will feel interested in one,—object to be beloved & played with—better than a dog anyhow.—Home, & someone to take care of house—Charms of music & female chit-chat.—These things good for one's health.—Forced to visit & receive relations but terrible loss of time.—

My God, it is intolerable to think of spending one's whole life, like a neuter bee, working, working, & nothing after all.—No, no won't do.—Imagine living all one's day solitarily in smoky dirty London House.—Only picture to yourself a nice soft wife on a sofa with good fire, & books & music perhaps—Compare this vision with the dingy reality of Grt. Marlbro' St.

Marry—Marry—Marry Q.E.D.

No children, (no second life), no one to care for one in old age.—What is the use of working without sympathy from near & dear friends—who are near & dear friends to the old, except relatives

Freedom to go where one liked—choice of Society & little of it.—Conversation of clever men at clubs—Not forced to visit relatives, & to bend in every trifle—to have the expense & anxiety of children—perhaps quarrelling—Loss of time.—cannot read in the Evenings—fatness & idleness—Anxiety & responsibility—less money for books &c—if many children forced to gain one's bread.—(But then it is very bad for one's health to work too much)

Perhaps my wife won't like London; then the sentence is banishment & degradation into indolent, idle fool—

Darwin's list of the reasons for and against marriage. Darwin was an inveterate lister and scribbler throughout his life—his twenty books being the mere tip of the iceberg of his handwritten jottings. The surviving correspondence numbers in the thousands of pages.

Charles with William Darwin, the Darwins' first child. By the time William was born in 1839, Charles was already convinced that life—including human life—had evolved. It was natural for him to compare the developmental behavior of William and his sister Anne with the recently arrived orangutan at the Regent's Park Zoo in London.

Anne, the Darwins' second child. On her death at age ten, Darwin apparently lost most of what was left of his religious faith.

Annie's box. This poignant collection of mementos and memorials to Annie, kept privately in her writing box by her mother, is a deeply treasured family heirloom.

Down House in Kent from the rear as it appears today, preserved essentially unchanged since Darwin's day.

Darwin's famous study in Down House. Darwin learned the importance of tidy organization while in his cramped quarters on the *Beagle*. The shelves above his desk on the right at the back each contained notes for an individual topic or chapter of the *Origin of Species*, reflecting earlier organizational habits on the high seas.

The Darwin drawing room at Down House. Here Darwin and Emma regularly played backgammon, with Darwin keeping a list that showed he won slightly more games than Emma. The bassoon recalls the experiment to see if earthworms react to music—a somewhat whimsical inquiry undertaken toward the end of Darwin's life.

The Wedgwood vase in the dining room at Down House. Emma and Charles were both grandchildren of Josiah Wedgwood, a close relationship that troubled Charles when he considered the effects of inbreeding in nature.

The Down House greenhouse—site of Darwin's many experiments with plants. Some of the experiments have been re-created more recently by the botanist and historian David Kohn in this same greenhouse.

The church at Downe, where Darwin and his family attended services. Charles was to be buried in the churchyard here—until a flurry of politicking managed to have him interred instead at Westminster Abbey.

The Queen's Head, one of two local pubs in Downe that had been there long before the Darwins arrived in 1842. The village has remained remarkably little changed over the years.

The Sandwalk, built for Darwin at the rear of their property, where he took daily strolls. Here, in solitude and privacy, Darwin undoubtedly had many of his inspired insights. Far from the madding crowd of London, it was also a place to feel safe as he kept his evolutionary thoughts a secret until forced to reveal them in 1858.

Darwin's walking stick. While on the *Beagle*, Darwin was one of the most agile and athletic members of the entire crew. The walking stick came many years later, as he kept up his devotion to the natural world in his outdoor walks.

early 1970s. (Geneticists nowadays have far less trouble contemplating such rates of evolution between species, some even pronouncing them rather conservatively slow.)

Uniformitarianism vs. Catastrophism. So "scaling" matters greatly in analyzing continuity and discontinuity against the background of true evolutionary (geological) time in the context of evolutionary rates. Recall that Darwin was heavily influenced by the geologist Charles Lyell, whose notion of "uniformitarianism" is central to Darwin's evolutionary ideas. Some historians have seen Darwin as the Lyell of biology, doing for the history of life what Lyell did for the history of the earth. Scholars such as the paleontologist George Gaylord Simpson have discussed the various nuances of meaning in the term "uniformitarianism" (whether as used by Lyell or later geologists), but I think Gould captured its double-edged meaning best, and most succinctly when (in a paper written in 1965 as a graduate student) he focused on just two separate meanings of the term: "uniformitarianism" is both a recipe for the scientific study of the earth *and* an assertion of the characteristic rates and nature of change of the earth's surface through time.

Uniformitarianism is often rendered by the catch phrase: "the present is the key to the past." Lyell was elaborating on the work of the Scotsman James Hutton (and other early geologists like Nicolaus Steno) in saying that the processes we see going on around us today—rainfall, erosion, the action of winds, the occasional volcanic eruption or earthquake—were always at work, and are themselves both necessary and sufficient to explain all the changes on the surface of the globe since the earth's inception, given the passage of sufficiently long periods of time. Scientific laws are immutable, Lyell was saying; gravity and other basic physical forces and phenomena have always been operating, and we need only apply them to the rocks exposed on the surface to understand how mountains are formed and thick layers of sediments accumulate.

But Lyell's uniformitarianism was also an assertion that such processes characteristically—if not invariably—act in a slow, steady, and gradational manner (linking concepts of rate of change with issues of continuity/discontinuity). Here Lyell was combating (as had Hutton before him) the old geological school known as "catastrophism," and Lyell's Christian faith notwithstanding, its links with biblical or "Flood" (still with us today) geology. Flood geology has its roots at least in part in the work of the influential eighteenth-century German geologist Abraham Gottlob Werner. Werner thought that the earth was formed through a succession of sediments and chemically precipitated layers, the last being the sediments left over after the biblical Flood (where Noah saved all the earth's organisms by placing them in pairs on his

Ark). It was Hutton who showed that many of the rocks exposed near his native Edinburgh were cooled remnants of ancient lava flows and other, more deep-seated "igneous" melts.

But the Deluge was, literally, a catastrophe, and Lyell's second meaning of "uniformitarianism"—"gradualism"—was seen to be the rational antidote to all the excesses of catastrophism in general, up to and including Noah's Flood. We can understand the world in terms of processes now operating, Lyell was saying, and for the most part those processes operate in a slow, smooth, steady, gradational manner. Discontinuity had been effectively banished from the geological thinking to which Darwin himself had been originally exposed. Darwin instead incorporated Lyell's uniformitarian views into his own geological theorizing, despite the fact that his mentor Henslow, himself a catastrophist, warned Darwin not to believe everything he read in Lyell's book.

Extinction. Discreteness/continuity; catastrophism/Lyellian gradualism; fast rates/slow rates—all are tied in with issues of observational scale. In many senses they were all the same issue, which came to the fore in slightly different contexts as geologists and biologists tried to wrestle with patterns of variation and replacement of different (albeit closely "allied") animals and plants, both geographically and through time. If catastrophism was already yielding to uniformitarianism by the time Darwin came along, it was still very much the dominant position—at least in England—when it came to that mystery of mysteries: the replacement of old, extinct species by new, albeit similar ones, in the next higher levels of rock.

Extinction had come to be nearly universally admitted, though the amateur paleontologist, and third president of the United States, Thomas Jefferson, based his desire to see exploration of the western lands of North America at least in part on his wish to see mastodons, giant ground sloths, and other Ice Age mammals that he felt certain were still alive, though then known only from bones preserved in sedimentary deposits in the East. What caused extinction was another matter—whether it was environmental change, the whim of the Creator, or other, more biological causes proposed by Darwin and those who followed. But most naturalists by the 1830s conceded that extinction was a real phenomenon, with natural causes.

The great French zoologist Baron Georges Cuvier (one of the relatively few noblemen to survive the French Revolution with both head and career intact) was a very sophisticated early scientist, best remembered, perhaps, for his detailed work as an anatomist (he is the "father" of comparative anatomy). Unlike his fellow Frenchman

Jean-Baptiste Lamarck, a distinguished specialist on living invertebrates, however, Cuvier was steadfastly opposed to any notion of evolution. Cuvier, along with the geologist Alexandre Brongniart, had produced a detailed map of the layers of sedimentary rock (primarily limestone and gypsum beds) as they are developed in the Paris Basin—one of the first geological maps to be published.

Cuvier saw that entire biotas—mixed assemblages of animals and sometimes plants—lived in a region for a time, only eventually to disappear. They always seemed to vanish rather abruptly, and always to be replaced by another fauna, often with species somewhat similar to the ones that had preceded them. In his *Discours sur les révolutions de la surface du globe* (1812), Cuvier identified no fewer than thirty-three revolutions, all interpreted to be separate acts of creation after God had destroyed His previous Creation. Cuvier's was a profoundly catastrophist position, embracing all the elements of fauna and flora; extinctions were real, but caused by God, and so were the separate acts of creation as new fauna and flora appeared to replace the preceding life assemblage.

Baron Georges Cuvier. Writing before Darwin, Cuvier saw the history of life preserved in the fossil record as consisting of a series of extinctions of many species, followed by the appearance of entire new sets of replacement species. He attributed the appearance of the replacement species to separate acts of creation by God. Though his explanation of the phenomenon was religious rather than naturalistic, the patterns he saw were real—and have only recently begun to be explained in evolutionary and ecological terms. It was Darwin who prevented Cuvier's patterns from being accepted as empirically valid; as he wrote in a marginal note in his 1844 *Essay*: "If species really, after catastrophes, created in showers over world, my theory false."

Lyell did not see things that way. Focusing on the marine fossils of the Tertiary Period (the last 65 million years), Lyell saw a progressive accretion of modernism: the fossil shells of the Paleocene Epoch had a small percentage of today's modern shells in them; the Eocene had more—and so on up to the Holocene ("wholly Recent") with 100 percent of the modern fauna finally showing up. Species became extinct because they could not live under the changing environmental conditions of the world; but new ones represented piecemeal, separate acts of the Creator—or so thought Lyell until, many years later, he finally acceded to Darwin's arguments after the *Origin* was published.

Discontinuity was the conceptual bailiwick of creationists—those who believed that God giveth and taketh away species—even entire faunas and floras. Gradualism and continuity became the watchwords for a natural science of the earth, even though such early proponents as Lyell could not go all the way and embrace a natural explanation for the mystery of mysteries—the introduction of new species.

As to species themselves, the Reverend William Whewell, with whom Darwin had often dined while at Cambridge, perhaps said it best, epitomizing the canonical, creationist view in his *History of the Inductive Sciences* (1837): "Species have a real existence in nature, and a transition from one to another does not exist." This sentence, published just as Darwin was opening his first Transmutation Notebook, summarized prevailing wisdom at the time.

It has always been assumed that religious ideology—a version of the history of the earth, of life, and especially of mankind based strictly on biblical accounts—was what not only inspired the early creationist naturalists like Cuvier but led them in a sense to deny the evidence of their senses, in effect to ignore or distort what they actually observed in the natural world. Nothing could be further from the truth: we may know more details about the history of the earth and of life, and the very nature of the living world, now than was known back then. But the patterns that these early, careful observers saw were very real, even if interpreted through a Christian mind-set.

Consider, for example, that the very first jottings Darwin made on evolution after alighting from the *Beagle* back home in England reveal him to have been a saltationist!

Darwin's Evolution

Alfred Russel Wallace set out on his first expedition already convinced of evolution; his goal—realized years later, well into his second journey—was to come up with a cause for it. In contrast, Charles Robert Darwin was by all accounts (his, his family's, and friends', as well as the overwhelming judgment of historians) still very much in the creationist mold when he left on the *Beagle*. He made a few suggestive comments—like the one about the Galapagos tortoises and especially the mockingbirds quoted in chapter 2. And he met John Herschel in Cape Town. Nearly a decade later (1845), in the second edition of the *Journal* ("*Voyage of the* Beagle"), he mentions that his suspicions were actually aroused by the Galapagos finches—though still only hinting at evolution:

> Seeing this gradation and diversity of structure in one small, intimately related, group of birds, one might really fancy that from an original paucity of birds in this archipelago, one species had been taken and modified for different ends.

Darwin saw the clues: the patterns of geographic replacement of rhea species in South America, the geologic replacement of "closely allied" fossil mammals through time, and the tortoises and mockingbird species replacing one another on adjacent islands. At some point these impressions became crystallized, conscious realizations: "The line of argument often pursued throughout my theory is to establish a point as a probability through induction, and to apply it as hypothesis to other points to see if it will solve them," as he wrote in 1838 (Notebook D). In his very first post-*Beagle* jottings (the "Red Notebook," begun on the *Beagle* but completed, with his earliest true evolutionary thoughts, shortly after his return home, to nearly every historian's agreement), he reminds himself to check the literature for all the distributions of South American species—to build up more cases "through induction," but also to test to see whether the pattern holds in general; two sides of the same coin.

Darwin's early notebooks from 1837–39 are both revealing and frustrating. For the most part, they consist of an eclectic mélange of jottings but very little in the way of coherent patches of observation, speculation, theory—thought. Everyone's notes are like that: the ideas are far more complete (at least remembered to have been so) than the often meager texts reveal in retrospect. Darwin himself went back over his notebooks years later, mining them for insights and facts gleaned from the literature. Indeed, he ripped out whole sections—and it is only because of the patience of a dedicated group of historians that the material has been reassembled to as close an approximation of the original as possible.* Darwin took only bits and pieces—and even

*Gavin De Beer, the developmental evolutionary biologist, published a five-part series entitled *Darwin's Notebooks on Transmutation of Species* in 1960–61. A later, more complete and scholarly annotated volume appeared in 1987, edited by Paul H. Barrett, et al., the source on which I rely here. Sandra Herbert was responsible for the Red Notebook, David Kohn for the Transmutation Notebooks.

scribbled on the cover of the earliest, the Red Notebook, "nothing of interest"— because his early ideas contained in that notebook had radically changed by the time he was writing up his theory for publication.

But there is no slow dawning of the truth-through-induction to be found in these precious early notebooks. Early on, Darwin comes to refer to evolution as "my theory." So we will never see the actual moment—or even the precise phase of his life (though I believe he was still on the *Beagle*)—when he abandons the creationist story for a naturalistic one, a story of evolution through natural causes. Nor should there be any mistake about how Darwin himself saw things *vis-à-vis* the older, creationist story. Throughout these early notebooks, written in a state of great creative excitement and intensity between 1837 and 1839, one strong theme emerges. Darwin begins to think that there are only two ways of understanding and explaining the patterns he saw: Either the replacement of one species by another species (whether over space or through time) has a natural cause. Or the replacements are done by God. Repeatedly, Darwin asks himself rhetorically (and wholly privately—this was a big secret, after all) Why? *Why would God bother (for example) to replace one set of edentate (the sloths and armadillo group) mammals in South America with another set of edentates—kinds of animals only found in the Americas?* But if one sort of animal was replaced through a natural process, an *evolutionary* process, the mystery disappears: evolution of one species from another would automatically create closely similar, closely related groups of species . . . and one would *expect* entire groups to be located in particular sections of the globe. Compare that vision with the predictions of Noah's Ark foundering on Mount Ararat (as far as I know, Darwin never wrote such a speculation down, but I'm sure he thought of it), where all the organisms on earth would have to have disseminated from one spot. One would never predict such orderly patterns as Darwin saw in the distribution of similar species in space and time. Darwin saw that there was only one "alternate" theory—and that was the creationist story. And to him it no longer made sense.

Others had seen the same sorts of clues and patterns that he saw. And it is indeed true that some of them—such as Lamarck in France, and his followers, like Darwin's mentor Robert Grant in Edinburgh—had already concluded that evolution must have taken place. Yet the idea was still deeply suspect, and wholly unpopular, among the leading scientists and scholars in Great Britain when Darwin was young. So why was he convinced when others had not been? It is often said that Darwin succeeded where others had failed in convincing the world that life had evolved simply because he had a causal mechanism worked out: natural selection. But Darwin himself was far from figuring out natural selection when he became convinced of evolution. So, again, why Darwin?

The answer seems to lie in the fact, first, that Darwin was raised a Nonconformist; he had both a grandfather and a teacher who accepted evolution; and, second, that Darwin was an aggressive, competitive young man who wanted his cake (a livelihood provided by his father the old-fashioned way) and to eat it, too. In short, he was determined to take his place among the ranks of respected young men of science—men who were no longer strictly the products of the monied, privileged classes.

Hence "my theory" in this treasury of late 1830s notebooks. Darwin's "theory" changed quite dramatically over the two or three years he sketched out his thoughts. He started with what at least superficially looks like "saltationism," and went through a somewhat Lamarckian phase. After he discovered natural selection, "my theory" came to mean natural selection—and not the more general theory of evolution itself (though he was to write years later that the latter was the more important of the two).

Above all, as we shall soon see, it was a subtle switch in how Darwin actually approached his own thinking that underlay some of the major changes as his thought developed—from the "Red" and "Transmutation Notebooks" of 1837–39, through the *Sketch* of 1842, the *Essay* of 1844, on up through the mature work of the *Origin*, and on to later revisions and further thoughts in *The Descent of Man* (1871).

There is another point about Charles Robert Darwin that I think has not been stressed enough: he may not have been the first, but he was most certainly the last, single human being to achieve astonishingly deep mastery of an impressively wide array of biological and geological subject matter—what have come to be, in the growth of knowledge and proliferation of scientific subdisciplines, entirely separate disciplines and subdisciplines. His experiences on the *Beagle* (indeed, as a fledgling naturalist even prior to setting sail) took the form of firsthand observations, collecting, and, to some degree, analysis. The creative frenzy of the late 1830s back home marked a stage when his impressions emerged to him as conscious laws, subject to testing and further analysis: the conclusion that life has evolved, followed shortly thereafter by his formulation of the causal process of natural selection. And he juxtaposed his reflections and insatiable appetite for reading everything relevant he could get his hands on, and maintaining a voluminous correspondence with all sorts of specialists and experts in a variety of pursuits, with experimentation (in domestic breeding; but also in plant reproductive biology, earthworm behavior, etc.) and taxonomy (his famous work on barnacles, consuming eight years). The man had, as he admitted himself, an insatiable curiosity about virtually everything that might bear on the "mystery of mysteries."

The breadth and depth of Darwin's firsthand experience with so many disparate fields is indeed breathtaking. Throughout his life, Darwin considered himself first and foremost a geologist; his knowledge of the fossil record was vast—albeit he collected few fossils after his epochal adventures in South America. He was likewise an expert in the distribution of plants and animals over the globe—overall perhaps the most crucial source of evidence convincing him that life simply must have evolved. But he also mastered the intricacies of comparative anatomy; of systematics (the classification of animals and plants), developmental biology, and the intricacies of inheritance. Add to that his deep interest in behavior, especially though not exclusively of animals (and even plants—one of his books was devoted to the "movement and habits" of climbing plants), for Darwin was very interested in Man the animal—our connections to the great apes, and through them, all the rest of life. As Randal Keynes has documented so well in his book *Annie's Box*, Darwin's comparison of the behavior of Jenny the Ourang at the Regent's Park Zoo with the development of his first two children, William (born 1839) and Anne (born 1841), came in the same period of great creativity in the late 1830s while writing his "Transmutation Notebooks."

The point is simply that no one since Darwin can fairly be said to have mastered so many different fields, all of them totally relevant to understanding the evolutionary process. Such mastery is absolutely critical not only to understanding Darwin's personal achievements but to understanding what has happened—and why—in the history of evolutionary thinking ever since. At the very least, this means that what a paleontologist such as myself has to say about what Darwin wrote—and how his ideas changed in the thirty-five-odd years between getting the idea of evolution and his last major contribution to evolutionary thinking in 1871—will perforce differ rather markedly from what a modern geneticist might think. Or a systematist, a developmental biologist, or a biogeographer. And given the diversity of experiences, theoretical outlooks, and downright opinion (however well considered) within each discipline, the task of writing an objective history of evolutionary biology from Darwin on seems almost as daunting as achieving a truly lasting synthesis of the rich array of data and theory from all biological and relevant geological disciplines into a single, coherent, and lasting evolutionary theory.

Writing from the impassioned perspective of a late twentieth-/early twenty-first-century paleontologist who has spent his entire career trying to puzzle out the relevance of patterns in the fossil record to understanding how life evolves, Darwin is not a hapless mouthpiece of the emerging ambience of his times. Neither is he a man so beset by worries about the reaction to his shocking views that he is driven—consciously or unconsciously—to sugarcoat them as much as possible, even to the point of painting a distorted picture of the way things are. The picture of gradual progressive change Darwin painted in the *Origin* may well have been consistent with other themes in the culture at large. And it might have proved more palatable than change-through-revolution to some of his anticipated readers. But Darwin himself passionately believed it to be a true picture of Nature.

He was a stickler for the truth, and always wanted to get his facts right. He acknowledged there were no really good cases of finely ("insensibly" was the term in vogue) graded series of fossils—and presented reasons for that, as well as for the apparent stability that species show in the fossil record. I have an alternate explanation for Darwin's thirty-five-year intellectual journey in the world of evolutionary theory, a world that he more than anyone else actually invented. I'll summarize it here in good Darwinian fashion: as an informed speculation that can be "tested" by delving deeper into his actual writings. For I believe Darwin's dance with evolution reflected changes in the very way he conceptualized the world and approached the task of being a scientist. The crucial event was his formulation of natural selection in 1838.

In a nutshell, Darwin went through what can only be described as an "inductive phase." He saw the patterns of replacement over time, and down the length of South America—patterns where closely similar species seemed to replace one another. He finally realized their evident meaning when he encountered the pattern of inter-island diversity among the tortoises and mockingbirds on the Galapagos. He gets home, and within six months opens up his first "Transmutation Notebook," proceeding initially to review the cases that act as clues pointing to the very fact of evolution.

The botanist and historian David Kohn has correctly pointed out that Darwin was also, from word one in his notebooks, looking for a *mechanism* of evolution. For, as Darwin himself says in the Introduction to the first edition of the *Origin*:

> In considering the Origin of Species, it is quite conceivable that a naturalist, reflecting on the mutual affinities of organic beings, on their embryological

> relations, their geographical distribution, geological succession, and other such facts,* might come to the conclusion that each species had not been independently created, but had descended, like varieties, from other species. Nevertheless, such a conclusion, even if well founded, would be unsatisfactory, until it could be shown how the innumerable species inhabiting this world have been modified, so as to acquire that perfection of structure and coadaptation that most justly excites our admiration.

Darwin is saying that the full case for evolution is incomplete without a mechanism to explain it—and the explanation he had in mind, of course, was natural selection.

I think that the evolution of Darwin's evolution reflects a switch in stance, once he discovered the principle of natural selection and became convinced of its truth and generality. The natural course would be then to see whether the patterns in nature—geographic and geological replacement of species, patterns of "affinity" revealed in systematics, comparative anatomy, and embryology—could be re-derived on first principles from the action of natural selection as Darwin imagined it.

It is very much as if Darwin wanted his patterns—the very "facts" that led him to evolution in the first place—to be retroactively seen as *predictions* that would necessarily be true if evolution has occurred (his first thought); and especially as evolutionary change is generated by natural selection (the even stronger position). That would mean that additional cases drawn from nature would serve as a test, rather than being just another example in an inductive pile that leads to the conclusion that life has indeed evolved.

What I am proposing here is that Darwin, rather consciously, switched gears from a predominantly, more-or-less Baconian inductionism—which got him, as he himself saw it, to the conviction that life has evolved. But as he later said, this was not wholly satisfactory: he wanted to discover just how species change through time. Once grasped, natural selection became the tail wagging the dog in the sense that all the patterns Darwin originally saw were, in essence, reformulated as predicted outcomes of the selective process. And as we have already seen, the great mid-twentieth-century geneticist Theodosius Dobzhansky pointed to the various levels of organization of biological systems in looking for continuity and discontinuity. Natural selection produces

*I.e., the "patterns" or "clues" pointing to evolution.

patterns of continuous variation within populations, said Dobzhansky, a point that of course was not lost on Charles Darwin one hundred years earlier.

That's it: if life evolves through natural selection, then evolutionary change must be gradual and continuous. Long before he got around to writing the *Origin* in a competitive burst once Wallace's manuscript arrived, Darwin's patterns were transformed into a series of predictions—all of which involved continuity and gradual modification. Those old patterns that simply would not fit were not discarded; Darwin was far too intellectually honest for that. But they were discounted or otherwise explained away, as "difficulties of my theory" that needed to be acknowledged. But it must also be said that Darwin did not play exactly by the rules of the hypothetico-deductive method. You are supposed to "rightly reject" your hypothesis if the data don't agree with your predictions. Fortunately for us all, Darwin stuck to his guns: there were too many patterns that supported his idea of evolution through natural selection for him to despair completely over the fossil record. Yet he might have asked himself more pointedly *why* the patterns of the fossil record did not agree with his predictions. For it was the form of the prediction, rather than the data of the fossil record, that was faulty.

Darwin's switch to a hypothetico-deductive stance once his innate inductionism led him to the truth of evolution (no mean feat!) seems admirable, and certainly a good rhetorical and intellectual stance to take when making his case for evolution through natural selection. But it caused the empirically valid underpinnings of catastrophism to be swept under the rug. Thus the utterly gradual and continuous nature of evolutionary change was—erroneously—accepted as empirical fact by the majority of evolutionists ever since Darwin, and right up to the present day.

The Red and Transmutation Notebooks

Setting out to read Darwin's Transmutation Notebooks is a little like watching Darwin himself set sail on the *Beagle*—full of enthusiasm, yet not knowing exactly what to expect. Like Darwin, we approach the notebooks with something of a prepared mind. And if, by the end of the journey, we do not arrive home with a theory so powerful that it is destined to shake the world, we at least have a richer picture of the genesis of Darwin's thought and the early stages of its "evolution"—preparing us for the *Sketch* (1842), the *Essay* (1844), and ultimately his *Origin of Species*. The theoretical structure Darwin devised by 1859 was so powerful that not only did it convince the interested thinking world that life had evolved but it still influences the way evolutionary biologists approach their subject matter in ways most of them do not fully comprehend.

So let's "do a Darwin" and, our own notebooks at the ready, record what seems most interesting in these notebooks as he jotted his thoughts between March 1837 and sometime late in 1839. The series begins with the so-called Red Notebook (now housed at Down House) and continues through Notebooks B–E (the "Transmutation Notebooks" proper, now housed at Cambridge University Library). Notebook A is devoted strictly to geology; Notebook B was opened at the same time and is devoted (as are Notebooks C–E) strictly to evolutionary and related biological issues. Notebooks M and N (also at Cambridge) are a separate series opened when Darwin moved on to Notebook D in order to record his thoughts on the more metaphysical aspects of his fervid thinking in these two highly charged years of productivity. They include speculations on the relation of human beings to his general evolutionary thinking. (Darwin instantly saw that the concept of evolution pertains as much to human beings as it does to any other species.) I will confine my voyage through Darwin's early written explorations of his "theory" to the Red Notebook, and to Notebooks B–E.*

In about a year, Darwin completed the Red Notebook, begun in late spring 1836 as a series of more or less perfunctory recordings of latitude, longitude, and so forth, while still on the *Beagle*. Historians have mulled this over, and pretty much agree that the latter half of the Red Notebook, still unfilled, was picked up and used by Darwin in good "waste not, want not" fashion after he arrived home.

I had understood that the Red Notebook contained Darwin's very first jottings on his transmutation theory—beyond the few tantalizing musings in his *Beagle* notebooks. I had read the *Origin* several times over the past forty or so years, so felt that I knew his thinking pretty well. The first time I read the *Origin*, I took a copy to a lecture by Louis Leakey. I was an undergraduate, anxious to learn about evolution but worried, as well, that I would not be able to understand Darwin's Victorian English and presumably recondite science. I was early and Leakey was late, so I delved into Darwin. I was vastly relieved to discover my fears on both counts were groundless, though a bit surprised to find myself immersed in details of pigeon breeding a few pages into the book rather than the glories of the natural world I'd imagined from Darwin's famed voyages.

*I rely here on the transcriptions of these notebooks published in 1987 by Barrett, et al., and the relevant texts by Herbert (Red) and Kohn (Transmutation Notebooks).

But I was totally unprepared for what I found in the Red Notebook. After many pages largely devoted to geology, all of a sudden on page 127 he begins with none other than the key examples that led to the core patterns—the clues that led him to evolution in the first place:

> Speculate on neutral ground of 2. ostriches; bigger one encroaches smaller.—change not progressive: produced at one blow if one species altered Mem [Memo]: my idea of volcanic islands elevated. then peculiar plants created. if for such mere points; than any mountain, one is falsely less surprised at new creation for large. Australia's = if for volcanic island then for any spot of land.*

There are a few more thoughts along these lines, and then, beginning (p. 128) with an exhortation to himself to go to the literature and examine the geographic ranges of birds and "animals" (meaning presumably mammals in South America), we find the following remarkable passage on pages 129–30:

> Should urge that extinct llama owed its death not to change of circumstances; reversed argument. knowing it to be a desert. Tempted to believe animals created for a definite time:—not extinguished by changed circumstances.
>
> The same kind of relation that common ostrich bears to Petisse . . . extinct Guanaco to recent: in former case position, in latter time.(or changes consequent on lapse) being the relation—As in first cases distinct species inosculate, so we must believe ancient ones, therefore not *gradual* change or degeneration. from circumstances: if one species does change into another it must be per saltum—or species may perish. This inosculation (representation) [Darwin's own later insertion] of species important, each its own limit and represented—Chloe Creeper: Furnarius. Caracara Calandria: inosculation alone shows not gradation.

I nearly fell off my chair when I first read these words. I had no idea that Darwin saw that closely similar species which replace each other either over space (e.g., the two rheas or "ostriches": the "Petisse" is the smaller, or "Darwin's" rhea), or over time

*This and all succeeding Red and Transmutation Notebook quotes are taken from the 1987 edition of Barrett, et al. I have smoothed out some of the abbreviations and other editorial insertions of that edition. The page numbers refer to the notebooks themselves rather than the printed edition.

(his example of the llamas) do not tend to become more similar to one another as their limits converge. Rather, the two species replacing one another geographically, or in time, are as distinct from one another where their ranges meet as they are when separated by greater distances in space or in time. This pattern of abrupt replacement led Darwin to think at first that evolution must be rapid—*per saltum*, by jumps—rather than some more gradual process.

This of course is just the opposite of what Darwin is known for: an insistence upon gradualism. Had he stuck to these themes it is conceivable that I would not be writing this book, since much of my life's work has been devoted to reconciling the truth of the sorts of patterns of abrupt change that lead to notions of rapid evolution of species ("per saltum") with the later-developed, accepted Darwinian vision of adaptation through natural selection. And though his "bulldog" Huxley was to admonish Darwin many years later that Darwin does himself a disservice in rejecting the dictum *Natura non fecit saltum* ("Nature doesn't make leaps") so unreservedly, Darwin of course never entertained such views again—despite his admission in the sixth edition of the *Origin* of the great stability shown by many species in the fossil record. So Darwin left a niche open for some of us who have followed in his footsteps.

In the next passage he goes on to liken the cutting of an animal (whether by accident or naturally) in two, or the production of offspring by a hermaphrodite, as analogous to what happens when new species arise from ancestors: a relatively sudden severing results in two individuals where, a short while ago, there was but one. "In the same manner," he writes (RN, p. 133), extinction may not depend upon aspects of adaptation. "There is no more wonder in extinction of species than of individual."

Darwin was flirting in these passages with the idea that species might have innate lifespans, more or less like an individual organism. The idea resurfaces periodically but there is little reason to believe it (and in any case no imaginable genetic mechanism for it). Darwin later decided that species extinction is primarily a reflection of losing out in competition with more highly evolved species (another supposition that has not weathered the test of time very well). Extinction is, to pretty much everyone's evident satisfaction in early twenty-first-century evolutionary biology, overwhelmingly a matter of environmental disruption; environmental change (usually physically caused) that goes too far too fast, overwhelming the adaptations of existing species—often many different species more or less at the same time.

But what is truly arresting is the simple analogy Darwin makes between individual organisms and species. Darwin is saying that species have a real existence, and like individuals, they have lifetimes (whether or not they are internally regulated, controlled, and predetermined) and deaths—extinctions in the case of species. And species, like individual organisms, have births—whether by radically dividing one species into two (the only possibility admitted in these passages) or in some other fashion. This is what he means by the tantalizing word "inosculation": like "kissing," inosculating means touching, joining, separating. It is a fleeting connection (and, according to David Kohn, with a sense of reorganization), alluding to the non-gradational aspect of the two species as they approach and replace each other in space (the rheas) or through time (the guanacos or the edentates).

What is this but a way to deal with Reverend Whewell's exactly contemporaneous dictum that "species have a real existence in nature, and a connection between them does not exist"? Darwin is saying in these, his earliest passages on evolution, that species indeed have a real existence in nature, but a connection between them *does* exist. Over a century later it became fashionable to speak of "species as individuals"—a movement begun by the biologist and Darwin scholar Michael Ghiselin. Species indeed have more or less discrete births ("speciation"), histories, and, eventually and inevitably, deaths (extinctions). It is stunning that this idea is fully present in Darwin's earliest grapplings with evolution, reflecting his reactions both to prevailing wisdom and, I think, more importantly, to the fact that species do tend to convey the impression of great stability and discreteness when details of the variation throughout their distributions in space and time are examined closely.

Darwin then veers abruptly back to geology. But as if a nagging thought he hadn't gotten down back on page 130 needs to come out, there is a quick note on page 153, picking up with the rheas ("Avestruz"):

> When we see Avestruz two species. certainly different. not insensible change.—Yet one is urged to look for common parent? why should two of the most closely allied species occur in same country?

Darwin here sees the multiplicity of species. If all species merely replaced one another, descendant following ancestor, in a single series, we of course would not see all the tremendous variety (minimally 10,000,000 species on earth right now) of life's diversity. But it is also an important way to address the non-intergradational features of the two "closely allied" species. For if both were derived from a now extinct common ancestor, the differences between the two descendants are more easily seen as separate evolutionary pathways of change from that common, ancestral state.

Darwin is wrestling in these early passages with absolutely fundamental issues: what are species? and how do they change over space and through time? While the answers to these questions he toys with were to change as his thinking progressed, these early thoughts reveal an attention to the problems of meshing an idea of evolutionary change with both philosophical pronouncements and empirical observations that he himself made on the apparent stability of species.

Notebook B. The first of the Transmutation Notebooks was begun in either June or July 1837 (the latter date is Darwin's own best guess; the former, the estimate of David Kohn, who transcribed Notebooks B–E for the Barrett volume). He had begun to get results back from the various experts to whom he had sent specimens. Perhaps the most notable were from the ornithologist John Gould, who was telling him wonderful things, such as that the two rheas were indeed distinct species; so, too, were the mockingbirds from several different Galapagos islands. And the assortment of little black and greenish Galapagos birds, with their array of different beak shapes, was not a mélange of different sorts of birds but, rather, a series of twelve distinct species allied to finches.

Darwin opens Notebook B with a sort of "call and response" exploration of Erasmus Darwin's *Zoonomia*, his grandfather's work which had explored evolutionary ideas. He quotes or paraphrases a topical snippet or theme from *Zoonomia* and proceeds to write his own thoughts on it. He begins with reproduction and variation, and on page 5 he writes: "With this tendency to vary by generation, why are species constant over whole country?" thus initiating a tension between apparent constancy of species and the finer-level variation that is always there within species.*

On pages 6–8, Darwin writes:

> Aegyptian cats & dogs ibis same as formerly but separate a pair and place them on fresh island. it is very doubtful whether they would remain constant

*This tension has never gone away since it is an ineluctable fact of nature's organization. It was the main focus of a team of geneticists and paleontologists I have worked with over the past few years, as we explored how it could be that species remain so stable for so long, given the prodigious amount of genetic variation typically present in local populations of the species. More on this later.

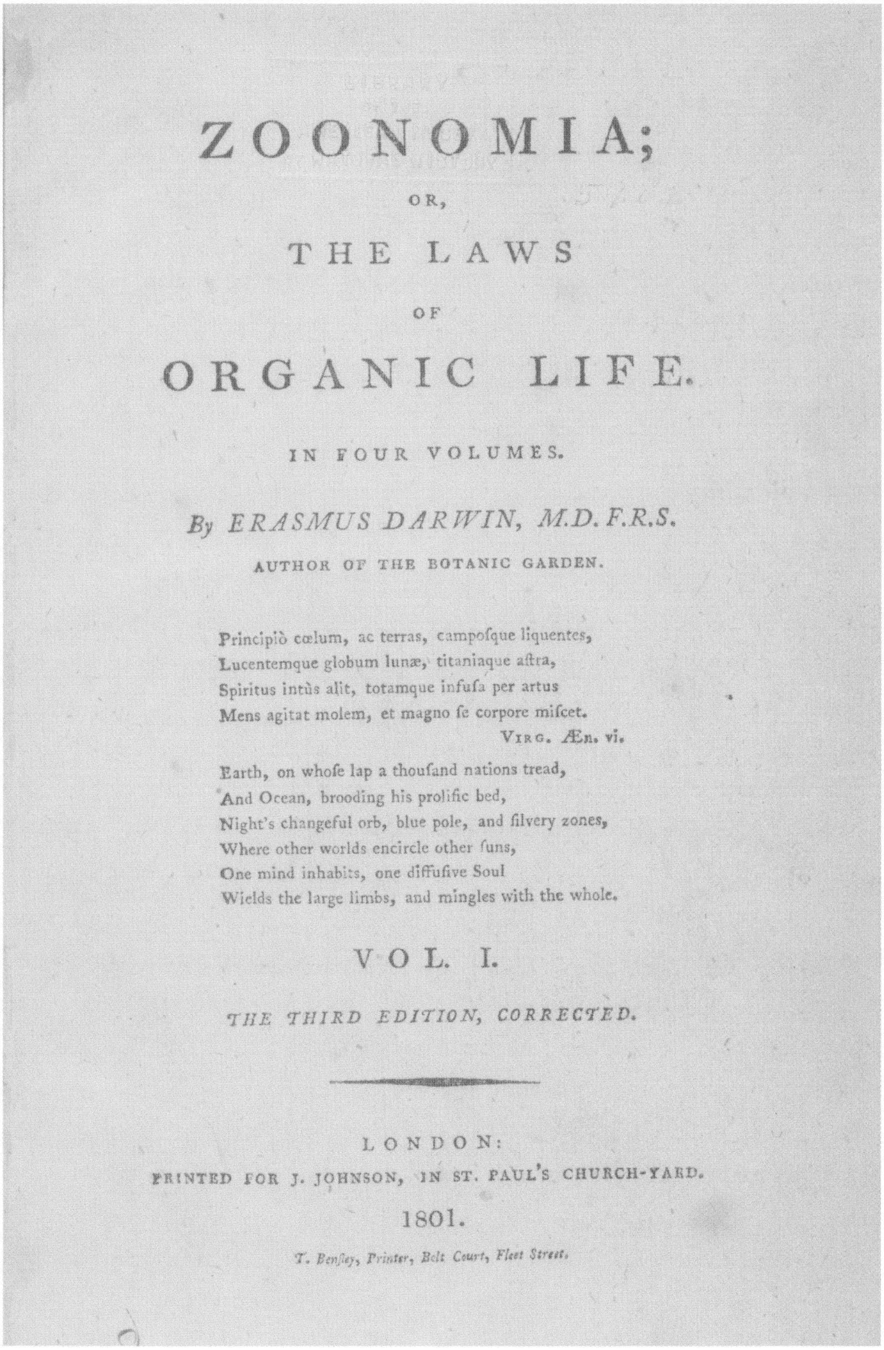

ZOONOMIA;

OR,

THE LAWS

OF

ORGANIC LIFE.

IN FOUR VOLUMES.

By *ERASMUS DARWIN,* *M.D. F.R.S.*

AUTHOR OF THE BOTANIC GARDEN.

Principiò cœlum, ac terras, campoſque liquentes,
Lucentemque globum lunæ, titaniaque aſtra,
Spiritus intùs alit, totamque infuſa per artus
Mens agitat molem, et magno ſe corpore miſcet.
VIRG. ÆN. vi.

Earth, on whoſe lap a thouſand nations tread,
And Ocean, brooding his prolific bed,
Night's changeful orb, blue pole, and ſilvery zones,
Where other worlds encircle other ſuns,
One mind inhabits, one diffuſive Soul
Wields the large limbs, and mingles with the whole.

VOL. I.

THE THIRD EDITION, CORRECTED.

LONDON:
PRINTED FOR J. JOHNSON, IN ST. PAUL'S CHURCH-YARD.
1801.

T. Benſley, Printer, Bolt Court, Fleet Street.

The title page of Erasmus Darwin's *Zoonomia*. The notion of evolution had been put forward by many people prior to Darwin, but no one had compiled the evidence or discovered the mechanism (natural selection) that provided the firm scientific basis for the concept.

> . . . according to this view, on separate islands, ought to become different if kept long enough apart, with slightly different circumstances; Now Galapagos tortoises, Mocking Birds; Falkland Fox—Chiloe, fox—Inglish and Irish Hare. As we thus believe species vary, in changing climate we ought to find representative species; this we do in South America closely approaching—but as they inosculate [i.e., meet, but two different, non-intergrading species replacing one another], we must suppose the change is effected at once—something like a variety is produced. . . .

Darwin is still thinking that patterns of abrupt replacement of one species by another one clearly "allied" to it bespeak a sudden process underlying the differentiation of the two. The problem of deriving new species on uninterrupted stretches of land (such as continental South America) contrasts with the relative ease of imagining evolutionary divergence when a male and female manage to get onto a fresh volcanic island. Isolation is an extremely important theme in Darwin's thinking—and, though it has had a spotty track record, it is still considered by most evolutionary biologists as key to both the generation and conservation of evolutionary change. Indeed, most of the battles in evolutionary biology that have been fought over the years were between two camps, one of which saw isolation as a prime ingredient in the origin of new species, and the second of which focused instead on processes of genetic change (mutation and other physiological genetic changes) as well as on such processes as genetic drift and, especially, natural selection, that modify gene frequencies in populations. And sometimes the debate has been whether isolation or natural selection is more important in evolution, as we shall see in chapter 5.

So Darwin in effect wonders how the isolation so readily seen on island systems like the Galapagos can occur in vast connected tracts of habitat like the Pampas of South America. On page 16, he goes on:

> I look at two ostriches as strong argument of possibility of such change—as we see them in space so might they in time. As I said before, *isolate* species & give even less change especially with some change probably vary quicker. Unknown causes of change. Volcanic island—electricity Each species changes, does it progress. Man gains ideas.

Much of this passage is unintelligible to me, but I find it fascinating as the typical jottings of a fertile, active mind exploring all sorts of peripheral thoughts suggested by the main topic. Everyone's notes look like this. But it is also important to watch Darwin speculate that what we see with the inosculating rheas in space probably also happens in time (possibly not a third rhea, the common ancestor in this scenario; rather, one rhea gives rise to the other, while itself persisting). And of course there is the notion of "progress" here as well, up to and including "man gains ideas"—progress, human progress, progress he himself is experiencing as *he* gains ideas. Heady stuff, all in a few sentences. Small wonder Darwin opened up a parallel series of Notebooks (M and N) to contain insights such as the latter ones in this paragraph that were far too much to cram into one focused notebook on transmutation!

Then Darwin discovers the metaphor of the "tree of life." It is the only coherent diagram (see Frontispiece and p. 109) in his Transmutation Notebooks, though there are some preliminary scratchings; his geology notebooks, in contrast, were loaded with cross sections and other sketches. Darwin derives the tree from looking once again at extinct South American mammals (p. 20):

> We may look at Megatheria, armadillos and sloths as all offsprings of some older type some of the branches dying out—with this tendency to change (& to multiplications when isolated, requires deaths of species to keep numbers **of forms** equable **but is there any reason for supposing number of forms equable; this being due to subdivisions and amount of differences, so forms would be about equally numerous.*** changes not result of will of animal, but law of adaptation as much as acid and alkali—organized beings represent a tree, *irregularly branched* some branches far more branched—hence Genera—many terminal buds dying, as new ones generated. [And then Darwin repeats a comment made earlier in the Red Notebook] There is nothing stranger in the death of species than of individuals.

This is a rich, arresting passage, crammed with new ideas, new ways of thinking about evolution. Perhaps most important is that Darwin sees that more branches (species) coming off, as subsets of a main branch, are clusters of related species: gen-

*Note: Here and on, the boldface type indicates a later annotation by Darwin.

"The covers of two of Darwin's four Transmutation Notebooks." Notebook B is most famous for Darwin's drawing of the evolutionary tree (see Frontispiece and p. 109). Notebook D contains Darwin's first recorded thoughts on natural selection (see p. 158).

era. This is the first statement he makes of the form: If evolution (ancestry and descent, the generation of new species from old) is true, what would it look like? It would resemble a tree, with the branches traceable back to thicker, older branches, right back to the trunk. But looking at his first sketch—and reading Darwin's statement that "The tree of life should perhaps be called the coral of life, base of branches dead" (p. 25), we see that the width of lines is constant—not thickening toward the base into a "trunk," but more like a succession of individual species. So the diagram more closely resembles what is nowadays called a "network" than a tree.

Regardless of the precise visual imagery, Darwin's logic here is important. This is the very first time, early on in his written evolutionary explorations, that he asks of the very notion of "descent with modification": If evolution has happened, what sorts of patterns would it produce? There is no record of his musing earlier on what is long since known as the "phylogenetic system." But here we have the earliest, very clear-cut glimmerings of his insight that the hierarchical structure of the classification systems of Linnaeus and a few other pioneering naturalists of the previous century is the expected outcome of the evolutionary process.

This is the first of the patterns that Darwin derives on first principles. It is a mode of thinking that quickly comes to permeate his entire approach to his subject. In this passage, for the first (documentable) time, Darwin leaves the realm of intuitive grasp for a sort of dawning of comprehension of the significance of mountains of similar cases into a general law requiring explanation.* That was the way he did indeed work for the most part on the *Beagle*, until he gathered the strength and analytic insight to conclude that life had evolved. Once again, I have to stress that I have never encountered anyone so keenly aware of how his own mind worked than Charles Darwin.

But this passage with the tree of life turns the entire thought process around one hundred and eighty degrees. In asking, "What patterns would we expect to see if evolution is true?" Darwin, having gotten the overall idea, switches into full-blown modern hypothetico-deductive mode. Now we can see if evolution is true by generating evolutionary trees—and then checking if they hold up over time with the generation of new data. Indeed, this is one of evolution's grand predictions—one that has been repeatedly tested over and over again—and a cornerstone in the scientific corroboration of the very idea of evolution. We see, for example, the close anatomical similarities between humans and chimps, and predict that they should share more genes in common than either would with any other living species. Sure enough, when the means of sampling the genes of different species became available, the test was run. The results: humans and chimps share over 98 percent of their genes, more than either species shares with gorillas, monkeys, squirrels, rats, or redwood trees.

Darwin also uses the term "adaptation" here, and I am grateful for David Kohn's footnotes, which explain that adaptation is not the "will of the animal" (a backhand slap at Lamarck) but a physical law, the sense of the otherwise peculiar allusion to acid and alkali—the latter terms (as Kohn puts it) "shades of Erasmus Darwin." We are still

*This is the form of intuitive, open-minded induction that critics (including myself once, along with Stephen Jay Gould, in a strongly worded passage in our original 1972 paper on "punctuated equilibria") erroneously state is psychologically impossible.

in the midst of the opening essay inspired by Darwin's rereading of his grandfather's work. But Darwin cannot specify, at this juncture, how adaptation does occur.

Yet there is one other bit that is at least equally arresting: Darwin maintains in the passage above that diversity of species is necessarily kept more or less equal through the ages. We now know that diversity has indeed gone through periods of such stability, but has also climbed rapidly in some phases in the history of life and declined precipitously in others. What is important here is that Darwin is willing to acknowledge that there might be some natural check limiting the number of species ("forms") that can co-exist at any one time.

I find this remarkable simply because Darwin had not yet read Malthus. When he did, he realized for the first time that there is a natural check on the numbers of individuals within all species—the lesson that allowed him to formulate natural selection in Notebook D. But here in Notebook B, Darwin is already exploring the possibility that there is a similar plateau, or average constancy, not for individuals within species but for the number of species that can exist at any one time. Because he quickly comes to the conclusion that extinction is largely the result of new species in effect outcompeting older ones, there is just a slight whiff of the sort of population dynamics (overproduction/competition) that leads to his formulation of natural selection among individuals within species. In other words, Darwin is, in a slightly different context, toying with the sorts of variables and modes of thinking that shortly thereafter enabled him to grasp natural selection.

Continuing the journey through Notebook B, we find Darwin contemplating—and answering—Georges Cuvier, the older French catastrophist, paleontologist, and comparative anatomist. After admonishing himself to "read his theory of the Earth attentively," we find the following call-and-response:

> Cuvier objects to (tran) propagation of species, by saying, why not have [sic] some intermediate forms been discovered. between palaeotherium, megalonyx, mastodon and the species now living—Now according to my view, in S. America parent of all armadillos might be brother to Megatherium—uncle now dead.

In other words, common ancestry, collateral descent, and extinction might be expected to explain what otherwise Darwin will see as a more or less common objection to "my theory," an objection that he himself will raise over and over again through the years as he engages in the well-known tactic of trying to anticipate the critics—beating them to the punch, blunting their rhetoric, and forestalling instant rejection because some obvious gaping hole in the theory was not at least acknowledged. Darwin was already bothered by gaps in the fossil record, an aspect of the radical reorganizations seen in both time and space—those "inosculations."

And he was moving toward seeing the evolutionary process as leaving "perfect gradations." On page 85 he writes:

> In some lower orders a perfect gradation can be found from forms marking good genera—by steps so insensible, that each is not more change that we know *varieties* can produce—Therefore all genera MAY have had intermediate steps—Quote in detail some good instances—But it is other question, whether there have existed *all* those intermediate steps especially in those classes where species not numerous (NB in those classes with few species greatest jumps strongest marked genera? Reptiles?) For instance there never may have been grade between pig & tapir, yet from some common progenitor—Now if the intermediate ranks had produced infinite species probably the series would have been more perfect, because in each there is a possibility of such organization.

At the very least Darwin no longer takes it for granted that the evolutionary progression between ancestral and descent species involves "inosculations" *per saltum*. He is toying with the thought that at least in speciose groups (groups with many species), there may well be more or less perfect gradations, although perhaps less so in groups with fewer species with greater gaps between them (i.e., in terms of their anatomical features). And for this latter case we have the possibility of common ancestry and extinction, yielding the sorts of missing data patterns that would explain these gaps. And (a few lines later), "this answers Cuvier."

Up to this point, Darwin occasionally uses the term "Creator" when he is clearly talking about the distribution of animals and plants and their changes over time and space. The following passage—by far the closest thing to a written, discursive passage on evolution that he had yet written—marks a watershed in his use of the term "Creator." This is a sort of parting of the ways, where Darwin comes out of the closet and, in effect, says that God may be the ultimate cause of all things, but He works through natural laws, including the natural laws that have been at work generating the entire history of life. This critical passage is, I feel, worth quoting in full (B 98–104), though it is the fourth paragraph that is most critical:

> The question if creative power acted at Galapagos it so acted that birds with plumage & tone of voice partly American North & South **& geographical distribution/division are not arbitrary & not permanent: this might be made very strong. If we believe the Creator creates by any laws, which I think is shown by the very facts of the Zoological character of these islands** so permanent a breath cannot reside in space before island existed—Such an influence Must exist in such spots. We know birds do arrive & seeds
>
> The same remarks applicable to fossil animals of same type, armadillo like covering created—passage for vertebrae in neck same cause, such beautiful adaptations yet other animals live so well.—This view of propagation gives. (no) hiding place for many unintelligible structures. it might have been of use in progenitor—or it may be of use—like Mammae on men's breasts—
>
> How does it come wandering birds. such sandpipers. not new at Galapagos—did the creative force know that these species could arrive—did it only create those kinds not so likely to wander. Did it create two species closely allied to Mus. coronata, but not coronata—We know that domestic animals vary in countries, without any assignable reason—
>
> Astronomers might formerly have said that God ordered each planet to move in its particular destiny—in same manner God orders each animal created with certain form in certain country, but how much more simple and sublime power let attraction act according to certain laws such are inevitable consequence let animal be created, then by the fixed laws of generation, such will be their successors—let the powers of transportal be such & so will be the form of one country to another—let geological changes go at such a rate, so will be the numbers & distribution of the species!!

Darwin himself becomes rightly enamored of this grand view, expanding the range of scientific law from the physical laws of motion applied to the external universe to the history of life on earth. It is a sort of battle cry—above all to himself—for now he simply must forge ahead and discover what those laws actually consist of.

Many years later, upon his death, his arch rival, the die-hard anti-evolutionist Richard Owen of the British Museum in London, said that biology still needed its Newton, but maybe Darwin took the rank in biology of a somewhat lesser physicist—someone like Laplace. Darwin himself would never have said anything along these lines. But he accepted his own challenge, and if there were to be a Newton of biology, it was indeed Charles Robert Darwin.

As Notebook B continues, Darwin starts to spend more space on patterns and mechanisms of heredity, beginning that search for an overarching law, a mechanism that can explain why organisms seem so well suited to their surroundings—their adaptations—and how those adaptations can be modified through time.

He gets surprisingly close, though seemingly without realizing it, right away. On page 146, he launches into calculations that have (even to modern eyes) some surprising results: "there will be a period though long distant, when of the present men (of all races) not more than a few will have successors. . . . Who can analyze causes, dislike to marriage, hereditary disease, effects of contagions & accidents." Darwin began these ruminative calculations by assuming that the population was constant, and then considered what happens if the population is increasing.

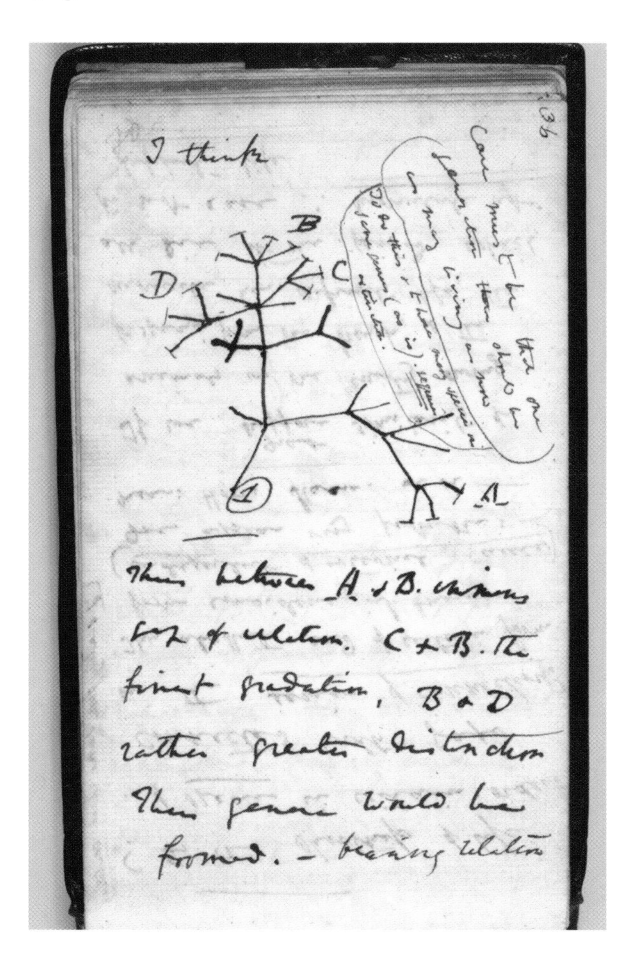

Darwin's sketch of an evolutionary tree of related organisms—the first of its kind, appearing on page 36 of Notebook B—scrawled after the dramatic words "I think" in 1837.

The usual point is that whole families disappear for a variety of accidental and more deterministic reasons. Darwin immediately leaps to the next level, saying that the *varying* races of mankind also suffer this kind of differential fate, then suggests that this is true for all animals, accounting for the extinction of some groups and the persistence of others.

With such thoughts and calculations, he is getting close to natural selection. He speaks about variation (albeit explicitly only among races of humans and by extension all other sorts of animals); he talks about random (accidental) and non-random causes taking out one person/race/species here and there, so that after a sufficient passage of time only a few of those present in the past have descendants now alive. To be sure, he does not link this to traits that may confer some differential benefits favoring persistence and continued reproduction. But at the very least, these pages show how familiar Darwin was with the general line of argument that is the real guts of the concept of natural selection.

Then he moves on, with yet another series of pages of notes of varying degrees of opacity—observations, notes on the literature, generally not linked explicitly to one or another critical issue. But now we come to page 209 and read:

> The reason why there is not perfect *gradation* of change in species, as physical changes are *gradual*, is this is after isolation (seed blown into desert) or separation by mountain chains &c the species have not been *much* altered they will cross (perhaps more fertility & so make that sudden step. species or not. . . .
>
> Why are species not formed. during ascent of mountain or approach of desert?—because the crossing of species less altered prevents the complete adaptation that would ensue

This sounds so modern! Darwin is saying that the isolation of parts of a species "not [yet] *much* altered" can lead to the emergence of new species. He goes further, actually invoking isolation as leading at least potentially to "abrupt divergence" (cf. his earlier use of the term "inosculation"), a view I find quite modern and attractive, though not perhaps the conventional view in modern evolutionary biology. Without isolation—as when looking at animals and plants distributed up a mountain slope or as a species range approaches the edge of a desert—there will be gradational variation, but new species will not emerge.

This is a fairly clear enunciation of what has long come to be known as "geographic" (or "allopatric") speciation, which is still very much a cornerstone of evolutionary biology. It does beg the question, what is a species? Fortunately, the question occurred to Darwin at this juncture and he immediately answered it:

> The passage in the last page explains that between Species from moderately distant countries, there is no test but generation . . . whether good species & hence the importance Naturalists attach to Geographical range of species—
>
> Definition of Species: one that remains at large with constant characters together with other (animals) beings of very near structure—Hence species may be good ones & differ scarcely in any external character. (p. 212)

Darwin here gives essentially what has been called the biological species concept, first articulated in a useful sense in modern evolutionary biology by Dobzhansky in 1935, though the idea is often associated with Ernst Mayr's writings in the 1940s. Species consist of individual organisms capable of interbreeding, sharing reproductive adaptations that allow individuals of opposite sex successfully to mate given the opportunity. Species are packages of genetic information—still very much the best way to think about them.

Species in a sense are self-defined as natural entities, packages of genetic information in nature. The problem is to recognize them when you see them—and that's precisely the problem Darwin sets forth in the beginning of this passage: how do we tell if two samples from "moderately distant countries" are distinct species or just different "varieties," i.e., variants within the same species? Especially because, as Darwin notes (using a hypothetical example of two virtually identical wren species living together that I have not included with the last passage), two species can be virtually identical but reproductively separate. Here the problem is how to be sure about the reproductive status of far-flung populations—a bugbear for those who are determined to assess as accurately as possible the "true" number of closely related species living in "moderately distant countries."

But, as Mayr pointed out in his *Systematics and the Origin of Species* (1942), the problem of telling true species from "mere" varieties in a geographic region is a bit of a godsend for an evolutionist. It is just this sort of spectrum between varieties and true species that supplies the gradational missing links that Darwin was beginning to wish for out loud (albeit strictly to himself) in his notebooks. It was a theme he was soon to discover and exploit.

For the moment, Darwin muses (p. 224) that if his theory is true, "we get a *horizontal* history of the earth within recent times"—an exceedingly subtle and sophisticated point. It is the reason that systematists can write a history of life on earth looking only at living species—a history that, of course, leaves out all extinct forms, but one that nonetheless captures the diversity and relationships of living beings with telling accuracy. These thoughts lead "you to believe the world older than geologists *think*." And it was Darwin the geologist, when he finally published his ideas, who did at least as much as anyone else in the mid-nineteenth century to establish the correct order of magnitude of geological time.

Where else might his "theory" lead? Why (pp. 226ff.) "to knowledge of what kinds of structure might pass into each other." And, "with belief of <change> transmutation & geographical grouping we are led to discover *causes* of change—the manner of adaptation. . . . My theory would give zest to recent & Fossil Comparative Anatomy, and it would lead to the study of instincts, hereditary. & mind hereditary, whole metaphysics . . ." and much more!

Darwin was nearing completion of Notebook B in early 1838. He wrote (p. 239) that "a gradual change can only be traced geologically (& then monuments imperfect) or horizontally & then cross-breeding prevents perfect change"—an early statement that the fossil record is imperfect, destroying remnants of "insensibly graded series," while interbreeding tends to smear out patterns of variation over space. He was already quite struck with the fact that interbreeding tends to counter patterns of geographic variation so that species tend to be "constant" sometimes over rather wide expanses. This latter tension between differentiation and homogenization—a problem addressed at the very outset in Wallace's 1858 manuscript sent to Darwin—is an important theme throughout the later notes, manuscripts, and, eventually, books. And Darwin's further allusion to a poor fossil record ("monuments imperfect") was destined to become a major component of the eventual presentation of his views—and whatever objections might be raised against them.

One final note: on page 278 (of a total of 280 pages), Darwin inexplicably writes: "In production of varieties is it not per saltum." The preceding page is blank. I have no idea why "per saltum" appears again since Darwin seems otherwise already well embarked either upon seeing gradualism or upon finding reasons why he does not. On the other hand, by March 1838, perhaps he had not yet fully decided to relinquish his first evolutionary scenario: the origin of new species (though here he says "varieties," even more surprising) *per saltum*—creating patterns of geographic and geological replacement; in other words, inosculation.

Notebook C. Notebook C was begun sometime in March 1838 and finished near the end of June of that same year. Much of it consists of discussions of other zoologists' travels, and aspects of the fate of what we would now call "genetic information" in the breeding process—as judged, of course, strictly through the heritability of physical traits of the organisms in question. In an interesting harbinger to Wallace's manuscript in which he argued that, unlike the case in domestic animals where "reversion to type" is common, there must be a force opposing such reversion among wild organisms, Darwin writes:

> Therefore the most hypothetical part of my theory, that two varieties of many ages standing, will not readily breed together: The argument must thus be taken, as in wild state . . . no animals VERY different will breed together.—We see even in domesticated varieties a tendency to go back to oldest race, which evidently is trending to same end, as the law of hibridity, namely the [pages not found, but seemingly followed on p. 33] animals unite, all the change that has been accumulated cannot be transmitted—hence the tendency to revert to parent forms, & greater fertility of hybrid & parent stock; than between two hybrids. (p. 30)

He goes on to speculate that long-entrenched varieties will produce hybrids between them, but not new varieties—a somewhat different tack from his previous thoughts on inosculations. Here he sees that "varieties" can become rather stable, in that way approaching the status of true species without yet having lost the capacity of their individuals to interbreed with those of other, closely related varieties.

Yet most of the early pages of Notebook C are crammed with specific mention of isolated facts and observations taken from the literature—with pithy summaries of aspects of his "theory" coming rather few and far between. Then on page 52 he once again links the notion of fertility (or rather, the lack thereof) with stability and change:

The botanist Joseph Hooker. Darwin took a liking to young Hooker, pumping him for botanical information and keeping up a lively correspondence. Hooker was one of the comparatively few scientists to visit Darwin at Down House. It was Hooker, along with Charles Lyell, who resolved the terrible dilemma brought on by the arrival of Wallace's letter and manuscript in 1858, although neither Hooker nor Lyell was convinced of evolution at the time.

not give yourself any trouble about them, for I know
how fully & worthily you are employed.
Besides a general interest about the Southern
lands, I have been now ever since my return
engaged in a very presumptuous work & which I
know no one individual who wd not say a very
foolish one.— I was so struck with distribution of
Galapagos organisms &c &c & with the character
of the American fossil mammifers, &c &c that I determined
to collect blindly every sort of fact, which cd bear
any way on what are species.— I have read heaps
of agricultural & horticultural books, & have never
ceased collecting facts— At last gleams of light have
come, & I am almost convinced (quite contrary
to opinion I started with) that species are not
(it is like confessing a murder) immutable. Heaven
forfend me from Lamarck nonsense of a "tendency to
progression" "adaptations from the slow willing of animals"
&c,— but the conclusions I am led to are not
widely different from his— though the means of change are
wholly so— I think I have found out (here's pre=
sumption!) the simple way by which species become

A portion of the letter written by Darwin to Hooker in 1844, in which Darwin alludes to the evolutionary clues he had seen a decade earlier on the *Beagle* and then, almost against his will ("it is like confessing a murder"), reveals his evolutionary views. Hooker and Lyell were the first people outside the family to be told of Darwin's conclusions.

Jean-Baptiste, Chevalier de Lamarck, the French scientist whose evolutionary views inspired Darwin's early teacher in Edinburgh, Robert Grant. Darwin himself paid close attention to Lamarck in his early days of evolutionary theorizing. But Lamarck did not address the details of the transformation of species, and Darwin had to develop his own ideas on the nature of the evolutionary process.

Thomas Henry Huxley. Darwin's "bulldog," Huxley was an energetic and highly visible lecturer—quite the opposite of the reclusive Charles Darwin. Huxley was much impressed by the evolutionary sequence of fossil horses he was shown when he visited O. C. Marsh's laboratory at Yale University.

The anatomist Richard Owen, who generally opposed Darwin's views. On Darwin's death, Owen commented that Darwin was no Newton but rather a lesser figure in the annals of biological science—presumably the remark of a jealous man.

The fossil *Glyptotherium*, a member of the glyptodont edentate mammals. Darwin collected pieces of the bony plates of glyptodonts, and realized they belong to the same basic group as the modern armadillos.

A modern armadillo, another edentate mammal. Darwin wondered: Why were extinct species replaced by other, similar species of the same group that lived nowhere else on earth?

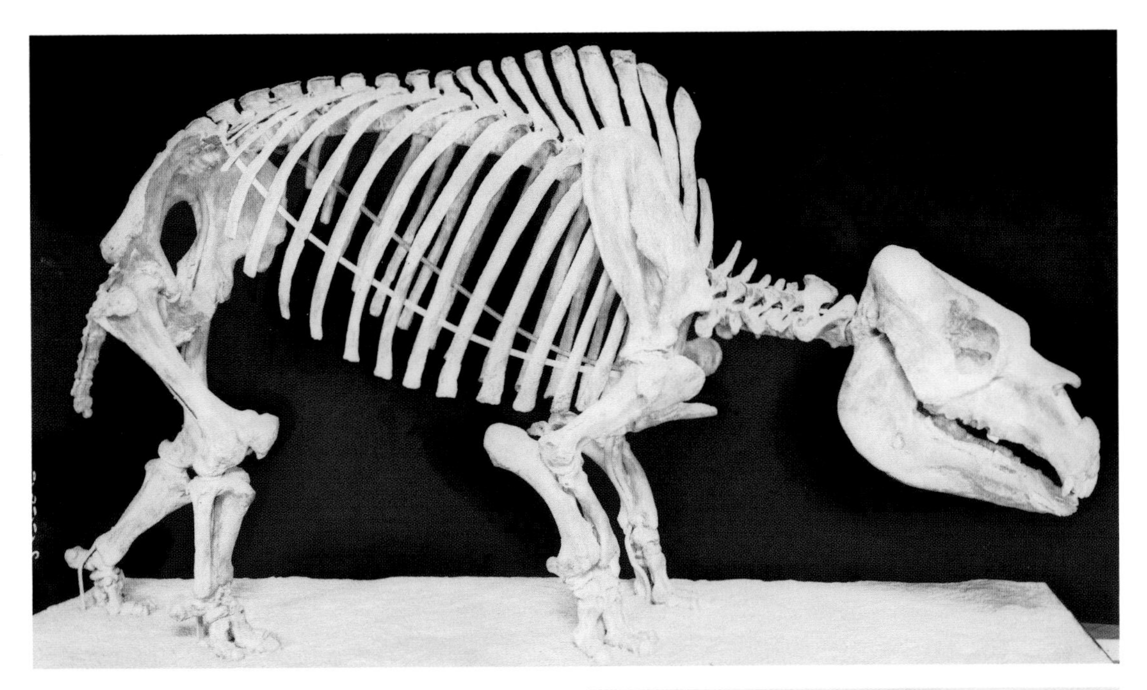

Toxodon, an extinct species of South American mammal. Darwin was erroneously led to believe that *Toxodon* was related to the capybara, the world's largest living rodent, and thus another example of replacement like the valid examples of the glyptodonts and sloths.

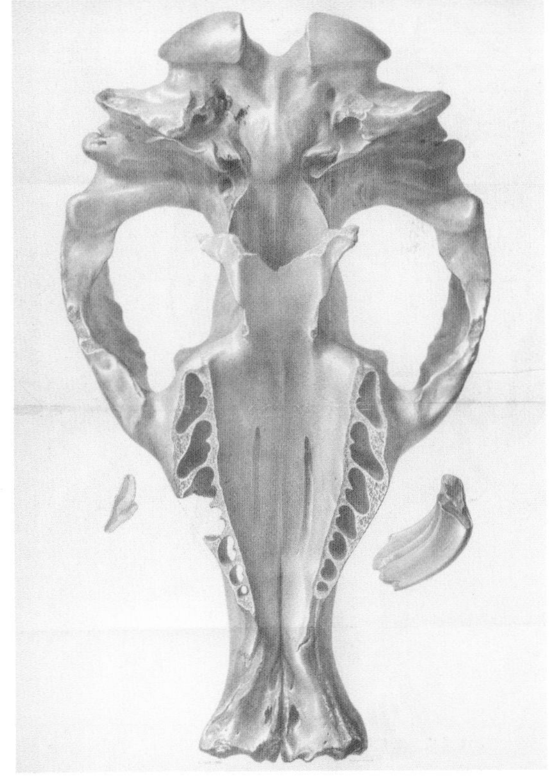

Drawing of the *Toxodon* skull that Darwin collected on the voyage. Richard Owen described *Toxodon* and had this figure drawn for his monograph on fossil mammals that were collected on the *Beagle* voyage.

Capybaras, the giant living South American rodent mistakenly seen by Owen as a replacement for the extinct *Toxodon*.

A guanaco, one of the species of South American camels, and a common sight for Darwin and his shipmates in southern South America.

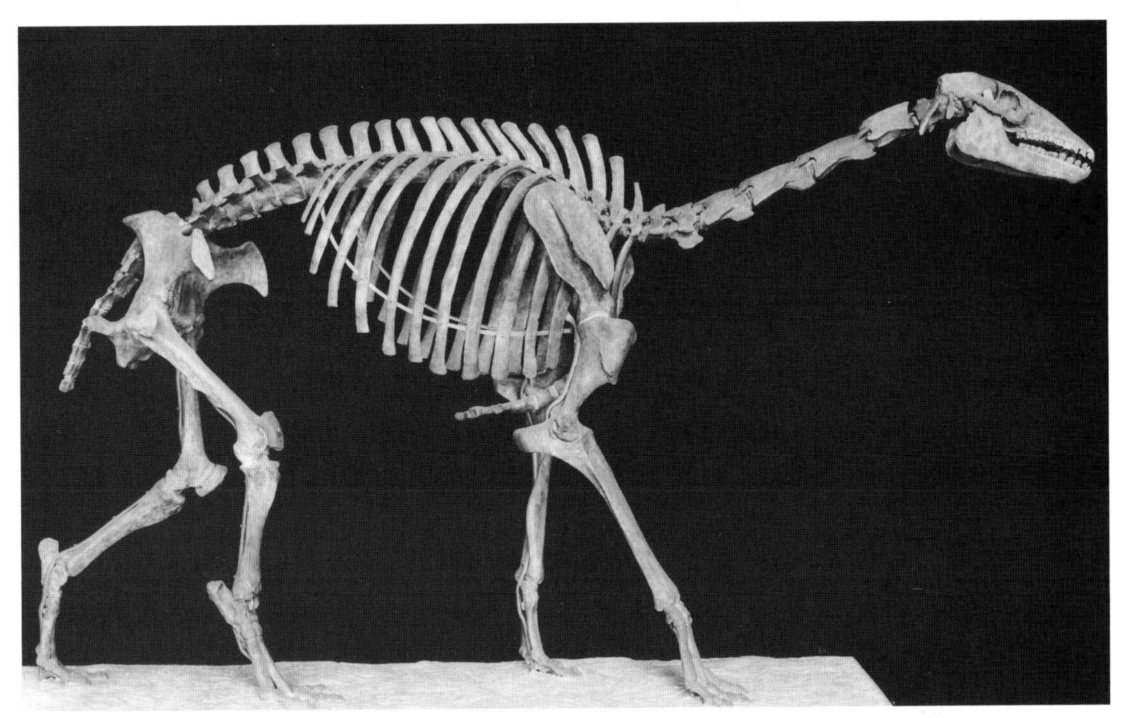

Macrauchenia, a camel-like South American extinct mammal. Darwin collected some bones of a *Macrauchenia* on the *Beagle* journey. Richard Owen concluded that *Macrauchenia* was an extinct relative of camels (hence llamas and guanacos), delighting Darwin, as it gave him another example of replacement of an extinct species by a closely "allied form" mentioned in his earliest evolutionary speculations in the Red Notebook. Later analysis showed that *Macrauchenia*, while superficially very camel-like (and thus an example of convergent evolution) was in fact a member of an extinct group of mammals known only from South America—and not related after all to guanacos.

Glossotherium, one of the extinct South American giant ground sloths. Darwin knew of the fossil ground sloth *Megatherium*, and thought that many of the fossil bones he collected on the *Beagle* were additional examples of these edentate mammals. Two species of sloths now live in South America. Again, Darwin wondered, why are edentates only here—and how are extinct species of sloths replaced by living ones?

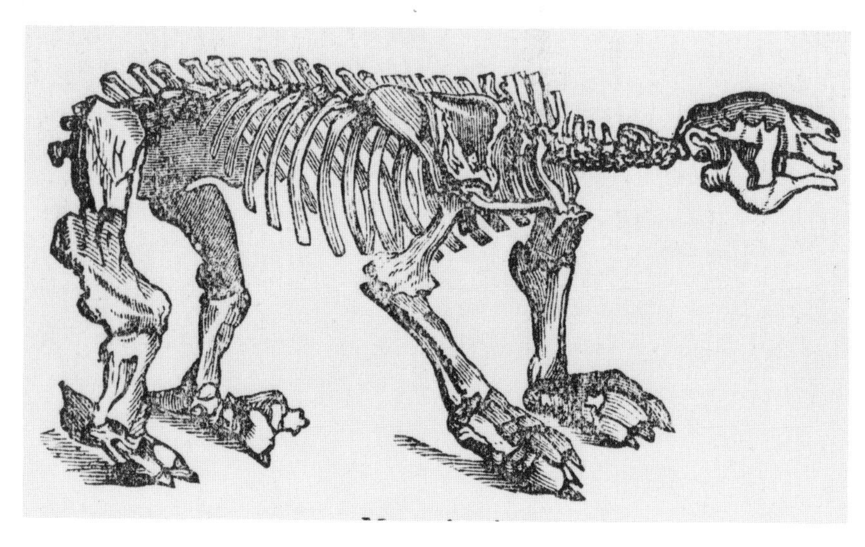

A contemporary rendering of the giant ground sloth *Megatherium* already known to the scientific world in Darwin's day.

A modern South American three-toed sloth. Thus two of Darwin's four original examples of replacement of extinct fossil forms by modern living species—belonging to groups that lived nowhere else in the world—were valid. From these he generalized, correctly predicting he would discover many more examples from around the world in the scientific literature.

> The infertility of crosse & cross, is method of nature to prevent the picking of monstrosities as Man does—One is tempted to exclaim that nature conscious of the principle of incessant change in her offspring, has invented all kinds of plan to insure stability; but isolate your species her plan is frustrated or rather a new principle is brought to bear. . . .

And every once in a while Darwin even has a minor "crisis of faith"—a statement of the other side's position, the stabilist/creationist position that he himself had started out with, sometimes felt so deeply that it takes something akin to a Gestalt switch to bring him back to the developing, in this case evolutionary, perspective. "Extreme difficulty of TRACING change of species to species <<although we see it affected>> tempts one to bring one back to separate creations—" (p. 64). He is thinking here of aberrations like ground woodpeckers. Once again, Darwin reveals that he always wanted to be the first to acknowledge the difficulties and criticism of his theory.

On page 70, he returns to—and makes even more explicit—the sliding scale of varieties/true species over a geographic landscape:

> Hence it is universally allowed that the discrimination of species is empirical. **show this by instances**—Once grant my theory, & the examination of species from distant countries., may give thread to conduct to laws of change of organization!

And on the pages immediately following, Darwin makes a rhetorically impassioned account linking "Man" with all the rest of Nature's productions—at one point comparing the behavior of orangutans (recently arrived at the Regent's Park Zoo) favorably with the behavior of naked wild savages.

But again, most of the pages of Notebook C are filled with notes and references to cases and examples drawn from the works of other naturalists, very much as if Darwin is out to take his patterns, his clues (drawn originally on but a few examples) not any longer by "induction" but by sheer enumeration, to make them airtight patterns pertaining to all organisms throughout the globe. Yet on page 106 he does make an interesting statement about the analogy between domestic and wild "varieties": "Two grand classes of varieties.; one where offspring picked, one where not.—the latter made by man and nature; but cannot be counteracted by Man. . . ." Still looking for his mechanism, Darwin here reveals his "prepared mind"—for natural selection, once discovered, will be doing the "picking" otherwise reserved in this passage for the actions of selective breeders.

The next arresting passage (p. 123) is not about Darwin's science per se, but rather its potential significance—*and what he ought to do about it*:

> Mention persecution of early Astronomers—then add chief good of individual scientific men is to push their science a few years in advance only of their age. (differently from literary men.—) must remember that if they *believe* & do not openly avow their belief, they do as much to retard, as those, whose opinion they believe have endeavoured to advance cause of truth.

It was an exhortation sadly unheeded. And it took a near calamity (Wallace's manuscript arriving in 1858, a full twenty years after this passage was written) for Darwin to have the courage to follow his own convictions on the value of advancing science by "a few years" by having sufficient faith in his own theories. These notebooks were kept, of course, in secret, and this passage can only be a private attempt to screw up his own courage.

The very next sentence (p. 124) is also somewhat off the subject of Darwin's development of his core ideas but a perfect example of yet another phenomenon of the creative mind caught in its most fertile moments. There are several passages throughout these Transmutation Notebooks where Darwin sees things about evolution that have either gone unexplored or that have taken decades, even centuries, for their importance to be acknowledged. And seldom (if ever) is Darwin given credit for them, since these are themes either never developed in his actual later publications or else overlooked in the otherwise voluminous piles of concepts routinely seen as emanating from Darwin's mind.

The case in point: "It is of the utmost importance to show that habits sometimes go before structures," meaning that behavioral changes often precede adaptive anatomical changes. This is a very real phenomenon, not given adequate attention perhaps until Ernst Mayr began exploring the theme in some detail in the 1970s.

Darwin continues to riff on, jumping in desultory fashion (to the reader's eye) from one familiar theme to another: the fact that hybrids form more readily between newly minted species than they do between species that have been longer separated (p. 135); then back to the point that the classification scheme reflects underlying descent, and the interesting observation that if fossil species had evolved more slowly, whether geologists might not find more evidence for gradual change. But on page 153, we find yet another passage which has elements of that elusive causal mechanism, natural selection:

> Changes in structure being <<necessarily>> excessively slow, they become firmly embedded in the constitution, which other marked difference in the varieties <<made by>> of Nature & Man.—The constitution being heredetary [sic] & fixed, certain physical changes at last become unfit, the animal cannot change quick enough & perishes.—

Quite apart from the use of the term "unfit," here we have a gross form of selection: extinction of an entire lineage as it fails to keep up with still further changing environments because the changes already accumulated become "heredetary & fixed." All Darwin had to do was apply this line of thinking on a finer scale, and talk about the relative survival of variants within each breeding population as time went by, and he would have had his natural selection. But he was getting close: natural selection was by now only a few months away.

On page 175, we find his first mention of the problem of complexity. It is the exquisite problem of the complexity of the human eye—and here Darwin simply acknowledges its presence, since he is as busy cataloguing counterarguments as he is in building an ironclad evidential case for evolution (all the while still on the lookout for the causes underlying it):

> We may never be able to trace the steps by which the organization of the eye, passed from simpler stage to more perfect. preserving its relations.—the wonderful power of adaptation given to organization.—This really perhaps greatest difficulty to whole theory.—[That Darwin follows this stark passage with "There is a breed of tailless cats near Bath, Lonsdale" conveys vividly the overall quixotic texture and flavor of these notebooks]

The remainder of Notebook C is the by now familiar mélange of notes on literature and ruminative speculations, none addressing explicitly, or at least *as* explicitly, the central themes of Darwin's theory as it then existed that are found in the passages quoted above. Darwin closes the notebook with a prodigious list of books "with reference to Species" that he has either already read or plans to. Lyell's *Principles of Geology* is of course on the list, as are, tellingly, Whewell's *History of the Inductive Sciences* and Herschel's *Preliminary Discourse on the Study of Natural Philosophy*. All three are down as already read—though Darwin makes no direct comment on any of them, but simply lists them along with many other more specialized titles.

Notebook D. Darwin began Notebook D in mid-July 1838 and finished it in late September or early October, only three months later. It is famous for its first statements of natural selection (beginning on p. 134 of a total of 180 pages). But, tempting as it is to race ahead to natural selection, we cannot neglect the gems strewn along the way before those critical passages appear.

The notebook opens with pages of disjunct observations much like the previous ones. Darwin does, though, focus on peculiarities of reproduction—and on the distribution of traits on various sorts of hybrid and otherwise unusual crossings. On page 15, for example, he considers hybrid sterility in mules. And on page 26 he exults: "Mine is a bold theory, which attempts to explain, or asserts to be explicable every instinct in animals."

Bolder still (pp. 36ff.) is his stark break from God the Creator in understanding the history of life:

> 16th August [1838]—What a magnificent view one can take of the world Astronomical <& unknown> causes, modified by unknown ones. cause changes in geography & changes of climate superadded to change of climate from physical causes.—these superinduce changes of form in the organic world, as adaptation. & these changing affect each other, & their bodies, by certain laws of harmony keep perfect in these themselves.—instincts alter, reason is formed, & the world peopled with Myriads of distinct forms from a period short of eternity to the present time, to the future—How far grander than idea from cramped imagination that God created (warring against those very laws he established in all organic nature) the Rhinoceros of Java & Sumatra, that since the time of the Silurian, he has made a long succession of vile Molluscous animals—How beneath the dignity of him, who is supposed to have said let there be light and there was light. . . .

It is worth remembering that this is not merely a slap at old, received Judeo-Christian cosmology. *It was the only available alternative theory to evolution*, and thus fair game to be compared with Darwin's own, ever more coherent views (here summarized as the changes wrought in organisms as they adapt to geographic and climatological change) and to each other—adumbrations of "co-evolution" in geological time. So, it represents a sweeping, albeit highly generalized summation of his evolution theory to that point. A theory, though, that still lacked a mechanism: how are those adaptations modified in response to the environmental changes that constantly occur?

Returning to familiar themes (pp. 51ff.), Darwin remarks once again that the natural scheme of classification reflects true affinity, in other words, what we would now call "phylogenetic" relationships. He then moves on (p. 53) to crystallize a theme present from his earliest jottings in the Red Notebook: that patterns of replacement of closely allied species look the same—and are to be understood in the same fashion—both over time and over space: "NB. How can local species as at Galapagos., be distinguished from temporal species as in two formations? by no way.?"

We are almost up to the promised land: Darwin's first inklings of natural selection. But I cannot resist singling out just one last intervening passage, one that once again shows us how Darwin, more than anyone else I can think of, was keenly, explicitly aware of the thought processes he was using. By now he is well into his analytic mode as he says (p. 67):

> Although no new fact be elicited by these speculation(s) even if partly true they are of the greatest service, towards the end of science, namely prediction.—till facts are grouped. & called. there can be no prediction.—The only advantage of discovering laws is to foretell what will happen & to see bearing of scattered facts.

Darwin was the consummate scientist of his day and a model for ours. He was open to intuitive/inductive patterns until he could crystallize and generalize those clues into laws; add more examples to strengthen their generality; and then see what other observations would follow as predictions that could then be used to test those very generalizations and related hypotheses. I should also say that by "prediction," Darwin was not referring to predicting the future course of evolution of this or that lineage. Rather, what should we predict we should find if that the notion of evolution is true?

And now at last we've reached that point in our trip through his notebooks—the "Galapagos point" in his jottings when he starts to develop the concept of natural selection. For after nearly seventy more pages of facts and speculations (mostly on heredity), and with the sort of laconic interjection of pure golden thought between much more prosaic entries, we read (p. 134):

> 28th [September 1838] I do not doubt, every one till he thinks deeply has assumed that increase of animals exactly proportional to the number that can live.—We ought to be far from wondering of changes in numbers of species, from changes in nature of locality. Even the energetic language of Malthus DeCandoelle* does not convey the warring of the species as inference from Malthus—increase of brutes must be prevented solely by positive checks. excepting that famine may stop desire. In Nature production does not increase, whilst no checks prevail, but the positive check of famine and consequently death.
>
> Population in increase at geometrical ratio in FAR SHORTER time than 25 years—yet until the one sentence of Malthus no one clearly perceived the great check amongst men—Even a few years plenty, makes population in Men increase, & an ordinary crop. causes a dearth then in Spring, like food used for other purposes as wheat for making brandy. Take Europe on an average, every species must have same number killed, year after year, by hawks. cold, &c—, even one species of hawk decreasing in number must effect [sic] instantaneously all the rest. One may say there is a force like a hundred thousand wedges trying [to] force into every kind of adapted structure into the gaps in the oeconomy of Nature, or rather forming gaps by thrusting out weaker ones. The final cause of all this wedgings [sic], must be to sort out proper structure & adapt it to change—to do that, for form, which Malthus shows, is the final effect (by means however of volition) of this populousness, on the energy of Man.

That's it. He finally has his long-sought mechanism. But there is no Eureka from Darwin, and he is soon back to recording breeding notes on mongrels, along with a host of other random-seeming facts and observations drawn from reading and from what people have been telling him. He doesn't take this newfound causal mechanism and run with it—not right away. But he must have spent hours thinking about it from the fall of 1838 onward. For when he opens Notebook E in October 1838, he goes right back to Malthus.

*Alphonse de Candolle (1806–1893), French botanist and biogeographer. Darwin misspells his name.

Notebook E. Darwin now has the mechanism within his grasp: natural selection (though apparently he did not call it that until the 1842 *Sketch*). The effect of this realization on his entire approach begins immediately in Notebook E, though the drama is somewhat muted by the reams of miscellaneous notes gleaned from his readings, and the new thoughts and directions consist of but a few short, very occasional passages. In that sense, Notebook E is much like its predecessors.

But here we watch Darwin turn things on their head—looking to see the implications of natural selection. For the first time he begins to entertain explicit thoughts on the nature and importance of variation: selection needs to act on differences among competing organisms within a species. Up to now many of his voluminous reflections have dwelled on patterns of inheritance (resemblance of offspring to grandparents, for example) rather than on variation—which, from his newfound Malthusian perspective assumes an importance he hadn't realized before.

He also now sees that the analogy between the handiwork of domestic breeders (later termed "artificial selection") and selection in the wild works very well as an illustrative device: what breeders can effect in a relatively few years time is impressive—and yet think how much more can be wrought by the analogous process, given the enormity of geological time. Indeed, Darwin marvels in several passages in Notebook E at just how prodigiously long geological time must be.

Beyond discovering whole new avenues to investigate, Darwin takes the first step toward what was soon to become his hallmark approach to the very "clues" ("patterns") that led him to adopt the idea of evolution in the first place. He began with them; but now, logically, he tries to derive them *ex post facto*, not just from the idea of "descent with modification" (what would we expect to see if evolution is "true"), *but from natural selection itself.* Evolution is already on the verge of becoming synonymous with natural selection in Darwin's mind. And here, in Notebook E, he takes that first step of looking at those original patterns anew.

The first pattern Darwin looks at in this new light is the pattern of change—or rather, the distressing lack thereof—as seen in the fossil record. Darwin, as we have seen, was already a bit worried about the stability of fossil species, as his perspective for other reasons shifted from saltationism toward a more gradual view. But natural selection seemed to him to demand a very slow and gradual pace of change—and here begins the famous series of worried notes about how the fossil record is so riddled with gaps ("like pages torn from a book," as he says in Notebook D) that its lack of concordance with his predictions flowing from natural selection can be safely ignored.

Begun in October 1838, and running to July 1839, the notebook was cannibalized by Darwin in late 1856. David Kohn reports that twenty-eight of the thirty-nine pages Darwin removed have now been found and incorporated into the published version.

Darwin begins on page 3 (the first two pages were never found) quoting Malthus to the effect that the causes of population increase and decrease are "as constant as the laws of nature." But he (Darwin) ". . . would apply it not only to population & depopulation, but extermination & production of new forms. . . ."

He then turns immediately to his now-deepened conviction of slow, gradual change—and also once again reveals his own keen awareness of how his mind is working (pp. 4–5):

> Octob. 4th. It cannot be objected to my theory, that the amount of change within historical times has been small*—because change in forms is solely adaptation of whole of one race to some change in circumstances; now we know how slowly and insensibly such changes are in progress. . . .
>
> Those who have studied history of the world most closely, & know the amounts of change now in progress, will be the last to object to this theory on the score of small change.—on the contrary islands separated with some animals, &c.—if the change could be shown to be more rapid, I should say there was some link in our chain of geological reasoning, extremely faulty.
>
> *The difficulty of multiplying effects & to conceive the result with that clearness of conviction, absolutely necessary as the foundation stone of further inductive reasoning is immense.* [italics added]
>
> It is curious that geology. by giving proper ideas of these subjects. should be *absolutely* necessary to arrive at right conclusion about species.

Thus Darwin knew what he was up to, beginning to reconceive his evolutionary patterns as expected outcomes of his ideas on evolution through natural selection.

*This is of course the problem posed by mummified and painted images of the fauna of ancient Egypt, already encountered in Notebook B, pp. 6–8.

He then looks at the fossil record a bit harder, remarking:

> My very theory requires each form to have lasted for its time: but we ought in same bed if very thick to find some change in upper & lower layers.—good objection *to my theory*: a modern bed at present might be very thick & yet have same fossils. does not Lonsdale know some case of change in vertical series. (p. 6)

In other words, Darwin expected species to remain more or less the same as long as external conditions (what we now often call "the environment") do not change and the species remains well adapted to the original conditions. But he acknowledges that in very thick beds, there really ought to be *some* change detectable between the fossils in the lowest beds and those at the top of the sequence. He is beginning to worry about this point.

On page 12, he notes that "Spaniards says [sic] *no Tortoises* in other places besides Galapagos"—harking back to one of his original patterns, though he initially had paid the tortoises little heed because he supposed they had been put there by sailors. But then he is back (p. 17) to the worrisome problem of the lack of observed change in fossil species—using his "leaves torn from a book" simile to blame the lack of observed change in fossil sequences on intermittent deposition of sedimentary layers:

> Species not being observed to change is very great difficulty in thick strata, can only be explained, by such strata being merely leaf, if one river did pour sediment in one spot, for many epochs—such changes would be observed.

A little later Darwin turns his attention to the new problem of variation, first speculating (p. 51) that his theory will lead to the discovery of laws of "co relation of parts, from the laws of one variation of one part affecting another." But then he immediately admits, parenthetically, that he himself is not up to the task: "(I from looking at all facts as inducing towards law of transmutation, cannot see the deductions which are possible)"—a statement which almost sounds as if his commitment to using the various patterns to establish the truth of evolution ("law of transmutation") precludes his use of deduction to establish laws of variation. But he has at least deduced from natural selection that variation is an important new subject he needs to consider.

Indeed, he goes on to say that variation is so important that the naming of new species from single specimens is "worse than useless"—though he undoubtedly came to feel otherwise in his later barnacle work. But he thinks that laws of variation, once worked out, will tell the naturalist what the limits of variation may be as a sort of an a priori guide to delimiting species, especially the pesky question of what is a variety and what a good species. Despite occasional pronouncements along these lines during the subsequent history of systematic biology, no such laws of variation have been discovered for the very good reason that none exist. As Darwin himself was later to acknowledge, some species are very variable, both within populations and between populations over space; others show very little variation and are hardly separable from closely related species.

On page 59, Darwin is elated: "Herschel calls the appearance of new species, the mystery of mysteries. & has grand passage upon problem! Hurrah." As we know, Darwin met with Herschel in 1836 in Cape Town and may well have already been exposed, if not to the phrase "mystery of mysteries" itself, at least to Herschel's openness to the search for a naturalistic explanation for "the replacement of extinct species by others." By the time he recorded this note, Darwin of course was well on his way toward achieving precisely that.

Then Darwin gets back to his main fear: that the facts of geology increasingly appear to him not to accord all that well with "my theory." Here he explicitly teases apart extinction from evolution: the one happening rapidly, like the extinction of Steller's sea cow, the "Lamantin" of the passage, and the other—evolution—happening slowly. Yet he continues to add notes to himself to keep thinking about the problem (p. 87):

> It must never be overlooked that the chronology of geology rests upon amount of physical change affecting whole bodies of species, & only secondarily, by assumption well grounded, on time;—therefore the mere loss of species, which may be the work of a few years as with the Lamantin of Steller tells much less (though it also the effect of change) than a slow gradation in form which must be the effect of slow change+ & Therefore precludes effects of catastrophes, which must serve to confound our chronology [CONSIDER ALL THIS] Extinction & transmutation, two foundations, hitherto confounded,. of geology.

This is an extremely important passage. Darwin is developing the argument that wholesale turnovers—used by geologists to establish geological time, as well as the empirical basis of Cuvier's interpretation of the history of life—have natural causes. But he sees these turnovers strictly as extinction phenomena: the death of species. They have nothing to do with transmutation proper, which happens much more slowly than the often near-instantaneous cases of extinction (like the seemingly isolated instance of the extinction of Steller's sea cow). Thus does Charles Darwin try to decouple the two ingredients of Herschel's "mystery of mysteries," the pattern of temporal "inosculation" that was arguably the first clue to present itself clearly as he contemplated the meaning of extinct and modern mammals in South America. He was to continue to entertain doubts that the turnovers of stratigraphic paleontology have nothing whatever to do with the generation of new replacement species. But nonetheless this was the view he in fact stuck with. In so doing, he let his commitment to gradualism override the empirical facts of the matter. It is the origin of an attitude that permeates down to the present day.

By March 12, 1839, Darwin's thoughts were back in the present, in a vivid description (p. 114) of how he saw the competitive struggle—within and between species—lurking just below the serene facade of nature: "It is difficult to believe in the dreadful but quiet war of organic beings. going on in the peaceful woods. & smiling fields." It was the poet Alfred Lord Tennyson who, in 1850, wrote the famous line, "Nature red in tooth and claw." Tennyson of course had not read these private words of Darwin, yet the world is not altogether wrong in attributing the genesis of Tennyson's thought to its proper source: Charles Robert Darwin.

Immediately thereafter, Darwin makes two very different analogies, connecting separate lines of thought he had already explored in earlier notes. The first (p. 117) connects his startling (and still underappreciated) calculations revealing how few progenitors have left progeny (in the human case, as little as five hundred years later) to the relationship of known fossil species to the species of the modern world:

> The theory that all animals have sprung from few stocks, does not bear, the least on ancient generic forms.—the animals in the Eocene period could not have been direct forebears of any of ours,—even if extinction is denied.

His second analogy (p. 118) is even more important, for this is where Darwin very briefly makes the connection between artificial selection by breeders and the natural process of selection in the wild—a theme that would come to dominate his entire approach to evolution as soon as he began to write formal discussions on the subject of "transmutation":

> Varieties are made in two ways—local varieties, when whole mass of species are subjected to some influence, & this would take place from changing country: but greyhound. & poutter Pidgeons (race horse). have not been thus produced, but by training, & crossing & keeping breed pure.—& so in plants *effectually* the offspring are picked & not allowed to cross.—Has nature any process analogous—if so she can produce great ends—But how.—. Make the difficulty apparent by cross-questioning—even if placed on Island—if &c &c.—Then give my theory—excellently true theory.

Darwin already had nature's "process analogous" firmly in mind, of course, and he admonishes himself essentially to argue how that process would work to make new varieties (and species) if members of a species invade an island—one way nature acts to restrict crossing the way breeders have learned to do. This passage is actually a recipe to re-derive from the principle of natural selection all the patterns of differentiation and replacement that had served as Darwin's original clues to evolution in the first place. He was to follow this plan throughout his writings, right up through the publication of the *Origin* over two decades later.

Shortly thereafter, he writes (p. 122, the section beginning "In the place" being a thought added later):

> The more I think, the more convinced I am, that *extinction* plays greater part then *transmutation*.—Do species *migrate & die* out.?—In the place where any species is most common, we need not look for change, because its number show it is perfectly adapted; it is where few stray ones are, that change may be anticipated, & this would look like fresh Creation. the gardener separates a plant he wishes to vary—domesticated animals tend to vary.

Here he is saying that extinction might happen more regularly than evolution—as species in the center of their range, where they are "perfectly adapted," will not in fact be expected to change unless and until conditions change. Rather, it is those members of a species that are few in number because they are occupying areas (usually near the edges of a species' range) which might be expected to show change, since their adaptations are not optimally suited for the environment they find themselves in. This is an almost astonishingly modern insight—yet another instance where Darwin's thought seems almost to have outstripped his ability to see all its implications. The lesson he draws here is that evolution is most likely to occur in marginal environments, and he makes the analogy with gardeners separating plants out. What Darwin doesn't see (as is all too clear from his subsequent writings) is that, in this passage, he holds the clue to understanding why species tend not to change: the real answer to his bugbear of non-transformation in fossils.

It was Darwin, arguably more than any of his contemporaries, who established the true magnitude of geological time. And though it is true that he did so simply to give his slow, gradual process of evolution the temporal room it needed to work, he derived his conclusions that the earth is very old from his own detailed experiences as a field geologist. Two passages toward the end of Notebook E show the argument developing:

> No one but a practised geologist can really comprehend how old the world is, as the measurements refer not to the revolutions of the sun & our lives, but to period necessary to form heaps of pebbles &c &c: the succession of organisms tell nothing about length of time, only order of succession. (p. 125)

Here Darwin neatly decouples the question of magnitude of geologic time from evolution because the evidence is purely geological, and not based on fossils, the potential charge of circular reasoning thus avoided. Later (p. 155) he returned to this theme:

> I utterly deny the right to argue against my theory, because it makes the world far *older* than what Geologists, think: it would be doing, what others but fifty years since to geologists.—& what is older—what relation in duration of a planet to our lives—Being myself a geologist, I have thus argued to myself, till I can honestly reject such false reasoning.

Just after the first of these two ringing assertions of the old age of the earth, Darwin makes yet another analogy, reverting to his "pages torn from a book," and not to the clue of species stability and peripheral evolution of a few pages back, to explain why the analogy doesn't really hold so far as the fossil record actually presents itself. On page 126 he waxes eloquent on how his old mentor Adam Sedgwick, writing with Roderick Impy Murchison, showed how the fossils of the Devonian Period were intermediate in aspect between those of the Silurian and Carboniferous Periods in a general sort of way ("without entering . . . into specific details"). There is general evolutionary connectedness among faunas. But then Darwin goes on to add: ". . . who can say how many centuries elapsed between each of these gaps, far more probably than during the deposition of the beds—The argument must be thus put, shall we give up whole system, of transmutation, or believe that time has been much greater, & that systems, are only leaves out of whole *volumes.*"

The final gems in Notebook E (save for the passage on geological time on p. 155 quoted above) come on pages 135–36. On page 135, Darwin sees that he has the explanation of stasis ("uniformity in geological formation") within his grasp, based on his thoughts that evolution should occur mainly where species are not well adapted to their surroundings. But he lets it go: "If separation in horizontal direction is far more important in making species, than time (as cause of change) which can hardly be believed, then, uniformity in geological formation intelligible." Sometime later he inserts: "No. but the wandering & separation of a few, probably would be most efficient in producing new species; also being reduced in numbers, but not so much these, because circumstances."

Darwin clearly sees the importance of isolation ("separation in horizontal direction") and that the natural version of selection will promote change in peripheral habitats—the essential ingredients of the theory of geographic ("allopatric") speciation. As we shall see, this strong statement becomes watered down in his published writings, leaving speciation theory itself on the periphery of evolutionary thinking until its resurrection in the 1930s and 1940s. Indeed, since then it has to some extent become once again muted.

But the inverse—that species in the core of their range will expectedly accrue little change so long as existing adaptations fit their living conditions—while tempting, was not to become part of his canon of thought. Nor, apparently, was his interesting (and also modern-sounding) speculation at the end of this passage: that populations reduced in size (so-called bottlenecks, in modern parlance) might also be expected to undergo evolutionary change.

We end this voyage through Darwin's Transmutation Notebooks with an explicit return to Malthus—Darwin's pronouncement of himself as a "Malthusian"—and his need to link variation in the wild with his Malthusian perspective (p. 137):

> It may be said that wild animals will vary, according to my Malthusian views, within certain limits, but beyond these not.—argue against this—analogy will certainly allow variation as much as the difference between species—for instance pidgeons—:then comes question of genera.

Darwin sees that he needs to document variation—and plenty of it—in the wild for his ideas to work. The question of variation is crucial: species to the casual eye seem so uniform, almost the equivalent of the apparent flatness of the earth to all but the most careful observer. Nor is the importance of variation just to serve as a groundmass on which selection works. Darwin needs to show that variation over space within species—sometimes sufficiently marked to lead to the recognition of geographic "varieties"—can mount up to be of the order of magnitude of the difference between species. He wants to see varieties as incipient species, and species as varying components of larger groups, like genera. And he wants to compare the parallel worlds of domestic breeding and differentiation and selection in the wild. He wants to paint a picture of gradations, the better to convince his peers that "transmutation" takes place and explains the entire history of life.

With this passage, we lose sight of Darwin's evolutionary thoughts until once again he surfaces with the exciting first attempt to write his theory out in a logical, readable, and (he hoped) persuasive form. He was obviously still reading and thinking about "my theory" between mid-1839 and 1842, although the birth of his first two children and the move to Down House were undoubtedly strong distractions. When next we pick up on Darwin's evolution, we find he has indeed progressed—in his compilation of examples, to be sure. But also in how he himself approaches his thinking. For by 1842, the revolution is complete. Switching cart and horse, Darwin starts with variation and selection and proceeds to use natural selection to re-derive afresh all those patterns he originally observed—the patterns that took him to evolution in the first place.

CHAPTER 4

Darwin's Evolution: The Manuscripts and Books

The 1842 *Sketch*—Part 1

Darwin's 1842 *Sketch* is to me more compelling than any of his other voluminous writings. It is a discursive discussion, a true essay—the first he ever wrote on "my theory." But it is also a coherent distillation of his Transmutation Notebooks, and as such continues to be filled with admonitions to himself to think harder about things. It is still very much a personal statement, written to challenge himself to organize his ideas into a logical framework. And, of course, it was written for Darwin's eyes only.

The 1842 *Sketch* consists of two untitled parts.* The first, with three chapters, has only eighteen pages. It is wholly given over to setting out Darwin's theory as it had by then taken shape: Chapter 1: On variation under domestication and on the principle of selection; chapter 2: On variation in a state of nature and on the natural means of selection; and chapter 3: On variation in instincts and other mental attributes.

*The version I am following here was published by Gavin De Beer in 1958. De Beer sticks closely to that originally published by Darwin's son Francis in 1909, and he has the good judgment to reprint Francis Darwin's extremely insightful Introduction to his own earlier editions of both the *Sketch* and the much-expanded *Essay*. The original *Sketch* consisted of thirty-five handwritten pages, in De Beer's edition coming to forty-eight printed pages.

Throughout his writings, Darwin was especially concerned to fold the evolution of behavior into his conceptual tent; he found it a daunting prospect to demonstrate to everyone's satisfaction that behavior was sufficiently supple and variable to be open to the process of natural selection.

It is in the first two of these three chapters that Darwin really lays his cards out on the table. The logical structure is clear—and sets the tone for all of his subsequent expositions, including the *Origin of Species*. He starts with the evidence that animals and plants vary—and do so in the wild every bit as much as they do in the barnyard and garden. In those early days, Darwin still thought that external conditions induced variation. The very first sentence reads:

> An individual organism placed under new conditions sometimes varies in a small degree and in very trifling respects such as stature, fatness, sometimes colour, health, habits in animals and probably disposition.

And he gets right to the point in dealing with domestic animals:

> Therefore if in any country or district all animals of one species be allowed freely to cross, any small tendency in them to vary will be constantly counteracted. . . . But if man selects then new races rapidly formed. . . .

Hence artificial selection, albeit not so named. He then asks if an analogous selective process can be found in nature. And of course he has already identified one. Far and away the most arresting item in these early pages is the simple subheading in chapter 2 consisting of the two words "natural selection." In the quasi-essay, quasi-notebook style that pervades this *Sketch*, Darwin writes:

> DeCandolle's war of nature—seeing contented face of nature—may be well at first doubted; we see it on borders of perpetual cold. But considering the enormous geometrical increase in every organism and as every country, in ordinary cases, must be stocked to full extent, reflection will show that this is the case. Malthus on man—in animals no moral restraint—they breed in time of year when provision most abundant, or season most favourable. . . .
>
> The unavoidable effect of this is that many of every species are destroyed

either in egg or young or mature. . . . In the course of a thousand generations infinitesimally small differences must inevitably tell. . . .

Nature's variation far less, but such selection far more rigid and scrutinizing.

This is by far his most coherent statement so far on natural selection. And in giving it a name, of course, he staked his priority on discovery of the process—though he chose snippets of the 1844 *Essay* rather than this *Sketch* to publish alongside Wallace's manuscript in 1858 to make certain his priority was recognized.

Alfred Russel Wallace in old age. Wallace independently derived the notion of natural selection while in Indonesia, twenty years after Darwin had discovered the same principle. But Darwin had delayed publication, and the arrival of Wallace's letter and manuscript in June 1858 was the crisis that finally forced him to publish without further delay.

Darwin then goes on to contrast natural selection with what he later explicitly named "sexual selection." Natural selection occurs because in the race for resources, or simply avoiding the natural perils of existence, the "variations" of some organisms work better—and thus increase their chances for survival and hence for leaving progeny. Sexual selection simply means that some organisms are better than others purely in the mating game. Darwin's distinction is utterly valid and useful—though it was to be largely ignored until the latter part of the twentieth century. He writes:

> Beside selection by death, in bisexual animals . . . the selection in time of fullest vigour, namely struggle of males. . . . Hence that male which at that time is in fullest vigour, or best armed with arms or ornaments of its species, will gain in hundreds of generations some small advantage and transmit such characters to its offspring. . . . (p. 48)

Nor did he stop with males. He saw that the same principle applied to females, who simply were better at the job of mothering than others in their peer group.

Darwin's summary of his theory in 1842 (pp. 57–58) is masterful, and worth quoting nearly in full. It contains a repetitious, almost chantlike exhortatory refrain: "and how can we doubt it." It focuses solely on variation and selection (leaving all the clues and patterns that brought him to the very idea of evolution in the first place to comprise Part 2), recast now as the observations that must be observed if his theory is true; or, in the case of the fossil record, the reasons why expectations are not met:

> *Summing up this division.* If variation be admitted to occur occasionally in some wild animals, and how can we doubt it, when we see thousands of organisms, for whatever use taken by man, do vary. If we admit such variations tend to be hereditary, and how can we doubt it when we remember resemblances of features and character—disease and monstrosities inherited and endless races produced (1200 cabbages). If we admit selection is steadily at work, and who will doubt it, when he considers amount of food on average fixed and reproductive powers act in geometrical ratio. If we admit that external conditions vary, as all geology proclaims they have done and are now doing—then, if no law of nature be opposed, there must occasionally be formed races, differing from the parent races. . . .
>
> . . . And therefore with the selecting power of nature, infinitely wise when compared with those of man, I conclude that it is impossible to say we know the limit of races, which would be true to their kind; if of different constitutions would probably be infertile with one another, and which might be adapted in the most singular and admirable manner, according to their wants, to external nature and to other surrounding organisms—such races would be species. But is there any evidence that such species have been thus produced, this is a question wholly independent of all previous points, and which on examination of the kingdom of nature we ought to answer one way or another.

So there it is: Darwin derives a theory for selection acting in nature, slowly transforming variations into races and, eventually, into species. He does not, at this juncture, mention isolation as the factor that ruptures the fertility among variants, and among races—producing true species that are (to use more modern phraseology) "reproductively isolated." But otherwise, this is a fair and ringing summation of his theory of the evolutionary process—the mechanisms of evolution—as of 1842.

And, of course, he finishes saying that we can go to nature to see if these predictions hold true. What he means by "examination of the kingdom of nature" is to look at the patterns and clues in the natural world—including, of course, the original ones that convinced him initially that life has evolved.

The *Sketch*—Part 2

Part 2 of the *Sketch* is also unnamed. Consisting of seven sections or mini-chapters (including a powerful summary statement at the end, chapters 4–10), Part 2 creates the template for all of Darwin's future discussions of the supporting lines of evidence for evolution, now converted into predictions (or problems) flowing from his "theory" of evolution through natural selection.

Darwin prefaces the first of these chapters with a general statement of logical principle: that, not knowing the will of God, it is impossible to predict any of the "relations" that such created organisms might have to each other, to geography, or even to time:

> But it would be marvellous if this scheme should be the same as would result from the descent of groups of organisms from the same parents . . . just attempted to be developed.
>
> With equal probability did old cosmogonists say fossils were created, as we now see them, with a false resemblance to living beings; what would the Astronomer say to the doctrine that the planets moved not according to the law of gravitation, but from the Creator having willed each particular planet in its particular orbit? I believe such a proposition (if we remove all prejudices) would be as legitimate as to admit that certain groups of living and extinct organisms, in their distribution, in their structure and in their relations one to another and to external conditions, agreed with the theory and showed signs of common descent, and yet were created distinct. (p. 59)

The only other explanation of general scientific observations is the old one of the Creator's will, which leads to no predictions, and which, in astronomy at least, has been abandoned when laws such as gravitation can be adduced to explain the phenomena. Why not then the same logical structure for the history of life?

First up are two sections on geology. (Francis Darwin says it is not clear where chapter 4 ends and 5 begins.) Darwin launches in with a fundamental prediction:

> Our theory requires a very gradual introduction of new forms, and extermination of the old (to which we shall revert). The extermination of the old may sometimes be rapid, but never the introduction. In the groups descended from common parent, our theory requires a perfect gradation not differing more than breeds of cattle, or potatoes, or cabbages in forms. . . . Now what evidence is there? (p. 60).

He goes on to develop a cogent argument explaining why the fossil record is so full of gaps: deposition occurs unevenly in different places, and in any case the probabilities of the preservation of animal and plant remains are always rather low. This, the forerunner of later, more detailed discussions in 1844 and in the *Origin*, is essentially the first modern statement of the discipline that has come to be known as taphonomy: the science of the formation and preservation of the fossil record. The statement contains much wisdom; the fossil record is indeed incomplete—and paleontologists then and now can only agree with Darwin that there is "no probability" that a perfect gradation would be found, even if all specimens that were preserved as fossils were collected and studied.

Yet, as in the earlier notebooks, Darwin says that, through time, specimens that are collected tend to link up "classes," and as such tend toward gradation. But he then says:

> Finally, if views of some geologists be correct, my theory must be given up. [Lyell's views, as far as they go, are in *favour*, but they go so little in favour and so much more is required, that it may be viewed as an objection.] If geology presents us with mere pages in chapters, towards end of a history, formed by tearing out bundles of leaves, and each page illustrating merely a small portion of the organisms of that time, the facts accord perfectly with my theory. (p. 63)

Though it is clear from the context that Darwin has in mind geologists who read the fossil record more literally than he can afford to do given the predictions he derives from his theory, we have to wait until the 1844 *Essay* for a more explicit statement of what such opposing views might consist of. Darwin does say, a few pages later, that "in older periods the forms *appear* to come in suddenly," but that's about all the time he devotes to alternative interpretations of the fossil record *vis-à-vis* his theory.

After taking another brief look at "Extermination" (Darwin says that extinction tends to be gradual, but not always so), he turns to "Geographical Distribution" (chapter 6). He begins by looking at the main divisions of life distributed over the earth's surface, then talks about how "chains of mountains, spaces of sea between islands and continents, even great rivers and deserts . . . form barriers of every kind to separate regions."

Then he gets to his predictions, starting with the "analogy of domesticated animals":

> Now, according to this analogy, change of external conditions, and isolation either by chance landing of a form on an island, or subsidence dividing a continent, or great chain of mountains, and the number of individuals not being numerous will best favour variation and selection. . . . Barrier would further act in preventing species formed in one part migrating to another part. Although no doubt change could be affected in same country without any barrier by long continued selection on one species.

Here Darwin is having it both ways: isolation by the imposition of barriers leads to the formation of separate species, though it remains a possibility that species can diverge by continued selection "in same country"—i.e., without barriers, as in the case of his South American rheas (as he saw them, at least). He sees the clear importance of isolation and in fact has a sense of the origin of reproductively isolated species through geographic isolation. But he clings to the notion that his theory does not fail in cases where isolating barriers are not readily apparent. The rheas and guanacos of the South American Pampas prevent him from ever insisting that the *only* way new species can arise is through a phase of geographic isolation.

There are more predictions:

> Now we can see at once that if two parts of a continent isolated, new species thus generated in them, would have closest affinities, like cattle in counties of England; if barrier afterwards destroyed one species might destroy the other or both keep their ground. (p. 69)

Thus there will be an expected tendency for species in neighboring areas to be more closely related to one another than to species living in more remote regions. And Darwin is aware that, once barriers are lifted, and new species come into contact with one another, one might drive the other to extinction—or they might "stand their ground," a possibility for explaining rhea and guanaco species distributions that in general has come to be the preferred view.

Darwin goes on to examine more specific predictions against the facts, but the main work of this brief discussion of geographical distribution is done. The predictions from his theory are clear-cut, and though migration and other factors may obscure traces of history, there is no doubt that the geographical distribution of animals and plants—unlike the geological distribution—accords beautifully with the expectations of his theory.

Next follow two closely related topics: "Affinities and Classification" (chapter 7) and "Unity [or similarity] of type in the great classes" (chapter 8). Darwin's discussion of "affinities and classification" follows beautifully from his discussion of geographic distributions: we see "the degrees of relationship are of different degrees and arbitrary—sub-genera—genera—sub-families, families, orders and classes and kingdoms. . . . If used in simple earnestness the natural system ought to be a genealogical one" (pp. 72–73). He then reiterates his theory on isolation and selection leading to the emergence of new species, and "thus we should get species of a sub-genus and genus—as varieties of merino sheep. . . ." Continued production of new species, in other words, creates clusters of closely related species (subgenera and genera), and as the process continues, still larger clusters of related groups of species are formed—the families, orders, and so on. Darwin saw that the Linnaean System—the *Systema Naturae* or "natural system"—was indeed natural, the obvious product of a long process of ancestry and descent. But here it becomes an expected, predicted outcome of his theory of evolutionary process rather than just one more valuable clue pointing to the very fact of evolution.

So too with "unity of type" (chapter 8), referring to cases such as "bat, horse, porpoise-fin, hand, all built on same structure." And here is his prediction: "But this unity of type necessarily follows on the theory of descent" (p. 76). He goes on to explain how descent entails modifications of existing structures, and hence a sort of family of divergent structures evolves from the original structure, yielding that "unity of type."

Of great interest is Darwin's extension of his evolutionary understanding of the facts of comparative anatomy to embryological development:

> This general unity of type in great groups of organisms (including, of course, these morphological cases) displays itself in a more striking manner in the stages through which the foetus passes. In an early stage, the wing of bat, hoof, hand, paddle are not to be distinguished. At a still earlier stage there is no difference between fish, bird, etc., and mammal. . . .

He adds that they can be distinguished, for example, by placement of the arteries. His point is rather that there is progressively more similarity the earlier in development embryos are compared—though he strikes a modern note in adding, "It is not true that one passes through the form of a lower group, though no doubt fish more nearly related to foetal state" (p. 78). In other words, comparative embryological development if anything enhances the "unity of type" of comparative anatomy—though we should avoid claiming that *Homo sapiens*, for example, always passes through a "fish stage" before getting to a higher vertebrate phase in development (an admonishment not always heeded by Darwin's intellectual descendants). And, of course, as is by now familiar, Darwin derives the fact that earlier stages of embryos of similar species (e.g., the vertebrates) resemble one another more closely than do the later stages directly as a prediction from his theory.

Chapter 9 is entitled "Abortive Organs." Here Darwin addresses what later came to be called "vestigial structures"—organs no longer used by certain kinds of animals or plants. The human appendix is an oft-cited (albeit arguably erroneous) example, as are the teeth of "rhinoceros. whale, narwhal," and other examples Darwin gives. He says in a note to himself that such structures are "eminently useful in classification. Embryonic state of organs. Rudiments of organs."

And, of course, such structures follow as an expected consequence of his theory. Darwin gives what is probably the first statement of preadaptation—or better, "exaptation"—when he says it is to be expected that organs evolved for a certain use ("adaptation") may come to be "turned to" another "purpose." But they might simply become useless, maintained only "by hereditary tendency."

And so he ends this section—and his review of predicted observations on his theory—by repeating his usual contrast with the received teachings of religiously imbued biology:

> I repeat, these wondrous facts, or parts created for no use in past and pres-

> ent time, all can by my theory receive simple explanation; or they receive none and we must be content with some empty metaphor, as that of de Candólle, who compares creation to a well-covered table, and says abortive organs may be compared to the dishes (some should be empty) placed symmetrically! (p. 83)

Darwin ends with a two-part "Recapitulation and Conclusion" (as chapter 10). His first section is sheer genius: harking back to the passage in Notebook D, page 36, he writes a mini-essay on Javan, Sumatran, and Indian rhinos, "three close neighbours, occupants of distinct but neighbouring districts, as a group having a different aspect from the rhinoceroses of Africa. . . ." He goes on to link these three species of rhino to all of his "predicted patterns": close affinity, unity of type, early developmental stages—and to the spectrum of variation bridging varieties and true species: "Now these three undoubted species scarcely differ more than breeds of cattle . . ." (p. 84). A fossil record of these rhinos is conveniently absent—and absent, as well, are any uniquely shared vestigial structures (though Darwin had earlier mentioned rhino teeth); otherwise he has a masterful summary of all his points with this one example of three Asian species of rhinoceros.

And so (what else?) he turns again to twit his favorite whipping boy, the only other game in town, creationism:

> Now the creationist believes these three rhinoceroses were created out of the dust of Java, Sumatra, these allied to past and present age and . . . with the stamp of inutility in some of their organs and conversion in others* with their deceptive appearance of true, not . . . relationship; as well can I believe the planets revolve in their present courses not from one law of gravity but from distinct volition of Creator. (p. 84)

Then he takes a further step: he tells us that "no naturalist pretends to give test from external characters on species"; rather, species are reproductively discrete entities. But are they always? ". . . sterility, though a usual, is not an invariable concomitant, it varies much in degree and has been shown to be probably dependent on causes closely analogous with those that make domesticated organisms sterile." There is a sliding

*Here he armwaves at geology, and manages to sneak abortive organs in after all!

scale of sterility, and in this way, Darwin manages to create a sliding scale between varieties (subdivisions of species based on external features and, generally, different locations within a species' range) and true species. Very clever—as he then goes on to say that, in such cases where two species can cross, a creationist must assume that two independent acts of creation can be made to blend into one another.

And so we come to the "Conclusion." It is in that somewhat grand rhetorical style that characterizes his later published summaries, not least, of course, the *Origin* of some seventeen years later. His patterns "cease to be metaphorical expressions and become intelligible facts. We no longer look on an animal as a savage does at a ship, or any great work of art, as a thing wholly beyond comprehension, but we feel far more interest in examining it" (p. 86).

Only geology emerges something of a loser, though even here there is a consolatory bone to throw: "Geology loses its glory from the imperfection of its archives, but how does it gain in the immensity of the periods of its formations and of the gaps separating these formations" (p. 86). He continues to question the "supposed creative spirit" as the creator of the myriad intricate patterns of geographic distribution and anatomical organization that flow out so simply from the natural evolutionary process he has described.

He ends* with a paragraph I shall quote in full. Readers familiar with the *Origin* will recognize the words:

> There is a simple grandeur in the view of life with its powers of growth, assimilation and reproduction, being originally breathed into matter under one or a few forms, and that whilst this our planet has gone circling on according to fixed laws, and land and water, in a cycle of change, have gone on replacing each other, that from so simple an origin, through the process of gradual selection of infinitesimal changes, endless forms most beautiful and most wonderful have been evolved. (p. 87)

These were virtually the same words with which Darwin was to conclude his *On the Origin of Species* when finally forced to publish his views.

*Well, not quite. This is a work in progress, and the final two paragraphs which follow are actually a telegraphic-style note to himself on what he needs to do to improve his arguments still further.

This brief, brilliant first essay—the real "abstract" of his views (Darwin himself considered the voluminous *Origin* as a mere abstract)—sets the content, tone, and very structure of both the 1844 *Essay* and the *Origin* as well. Creatively speaking, it is all over. To be sure, there are nuances of difference—changes in his views—and it will be worthwhile taking a brief look at the *Essay* as well as, of course, the *Origin* (first and sixth editions), in addition to *The Descent of Man*, to make note of his further "evolution" and to pick up some other insights into his thinking as well.

But 1842 is it: a distillation in mostly essay form that summarizes all those maddeningly brief and skittishly disposed insights of his Red and Transmutation Notebooks, while adding more examples and formulating additional insights. There are some more insights to come (the "principle of divergence" being perhaps the most important). But to a remarkable extent, Darwin's evolutionary theory is written for all to see (thanks to Francis Darwin and Gavin De Beer) in 1842.

The *Essay* of 1844

The 1844 *Essay* is much longer than its 1842 precursor—164 printed pages in the De Beer edition, compared with the 48 pages of the earlier *Sketch*. Yet in overall structure, the two are identical: same chapter heads, same division into two parts. The major difference is that in 1844, Darwin for the first time is writing not so much for himself but as if others will be reading his theory. The *Essay*, recall, was wrapped up and stored along with a covering note to Emma (with suggestions for suitable editors, and a financial provision of £400) requesting her to publish it in the event of his death. The expansion in length comes largely from the addition of more examples, rather than the introduction of entirely new subject matter, as well as from the verbosity that a true essay entails. He has left annotations, but otherwise has abandoned telegraphic passages here.

But this is not to say that there are no new directions of thought, and no surprises. Because the basic job was done in the 1842 *Sketch*, it is just these nuances and departures that deserve our attention as we take a quick look at the 1844 *Essay*, the *Origin*, and *Descent of Man*. One such nuance is a slight retreat from the more open, almost brazen approach of deriving pattern from "my theory"—the generation of variation and the action of natural selection. From 1844 onward, Darwin is more prone to say that the patterns of geography, comparative anatomy, and embryology are simply explained by his theory rather than the stronger form that they arise as simple predictions ("expectations") from his theory. This is not so much a change in logic as it is in rhetoric: from now on, Darwin is out to convince the reader of the truth of evolution—and that his theory of the mechanisms of evolution is correct. The result, inevitably, is an essay that seems less fresh, less fraught with creative excitement than can be glimpsed in the occasional passages of the Red and Transmutation Notebooks—and perhaps even more evident in their initial distillation in the *Sketch* of 1842.

And, as is apparent even in some of the titles and subtitles of his sections, Darwin's commitment to smooth transitions and gradual change is if anything heightened in this essay. For example: "Gradual appearance and disappearance of groups" and "On the graduated complexity in each great class."

As far as the core of the theory is concerned—the generation of heritable variation and the analogy between artificial and natural selection—there is little new in 1844. Darwin still thinks that much variation is induced by external "conditions of existence," especially as environments change. He continues to admit that there seems to be less variation in the natural state than is commonly seen in domesticated animals and plants. And, in what seems to me to be the greatest inconsistency in any of his writings, Darwin does spend considerable effort in the first portion of this *Essay* blurring the distinction between varieties and species. Though he agrees with Lamarck—to the effect that it is "seldom" difficult to tell species apart locally, the problems in deciding what is a species and what are merely "varieties" of species arising only over larger chunks of geography—he immediately launches into a list of examples where even locally the problems of species delimitation can be great.

Darwin even backs away in some places from a strictly reproductive concept of what species are: "The sterility of species, or of their offspring, when crossed has . . . received more attention than the uniformity in character of the individuals composing the species" (De Beer's edition, p. 123). There is, he says, an entire spectrum of degrees of infertility between species—another attempt to show gradation between differentiation within a species, and the formation of actual new species.

But then he seems to take it all back toward the end of the essay, now wholeheartedly agreeing with Lamarck that the distinction between varieties and species is blurred only when large geographic ranges are concerned—an observation Darwin says (p. 206) "seems chiefly to have urged Lamarck to the conclusion that species are mutable" in the first place. So there is a slight conflict between Darwin's presentation of the evidence supporting the mechanisms of evolution in the first section of this essay, and his discussion

John Gould's rendering of "Darwin's rhea"—forever known by that common name despite the fact that Alcide d'Orbigny had previously given the species another scientific name.

The greater, or common rhea. The two species of rhea met in only one place that Darwin knew of. Thus they appeared to "replace" one another geographically. Nor did they appear to blend into one another where they met. Darwin wondered, why would one species appear to replace another over space—both very similar, both coming from the same group of "allied forms"? This was the second of the three patterns of species replacements Darwin encountered on his *Beagle* voyage.

Some of the Galapagos Islands. Top to bottom: Charles Island; Chatham Island; "watery place"; and Albemarle Island. Charles and Chatham Islands are among the oldest of the archipelago; each has its own distinct species of mockingbird.

View from space of Fernandina to the left (west) of Isabela, the largest of the Galapagos Islands. The older islands lie farther to the southeast, and tend to have the better-differentiated species, such as the separate species of mockingbirds.

Darwin began to speculate that the tortoises from the different islands were well-differentiated members of the same ancestral stock when he was writing his notes on mockingbird variation—not long after the *Beagle* sailed from the Galapagos west into the Pacific.

The Galapagos mockingbird, one of four species in the Galapagos. Here was Darwin's third pattern leading him to the idea of evolution: Why did closely similar but distinct species replace one another on different closely spaced islands in a single archipelago? Darwin's notes on these species contain his earliest known writing on the possibility of evolution: "If there is the slightest foundation for these remarks to zoology of Archipelagoes—will be well worth examining; for such facts would undermine the stability of Species."

(15

Ornithology — Galapagos

3304 Gull : male
3305 Dove : do : One of the most numerous birds in the Islands.
3306 Thenca : male : Charles I^{d} —
3307 Do : Do : Chatham I^{d}. —

These birds are closely allied in appearance to the Thenca of Chile (2169) or Callandra of la Plata (1216). In their habits I cannot point out a single difference; — They are lively inquisitive, active, run fast, frequent houses to pick the meat of the Tortoise, which is hung up. — sing tolerably well; are said to build a simple open nest. — are very tame, a character in common with the other birds: I imagined however its note or cry was rather different from the Thenca of Chile? — Are very abundant, over the whole Island; are chiefly tempted up into the high & damp parts, by the houses & cleared ground.

I have specimens from four of the larger Islands; the two above enumerated, and (3349 : female. Albermale Is^{d}) & (3350 : male : James I^{d}). — The specimens from Chatham & Albermale I^{d} appear to be the same; but the other two are different. In each Is^{d}. each kind is exclusively found: habits of all are indistinguishable. When I recollect, the fact, that from the form of the body, shape of scales & general size, the Spaniards can at once pronounce, from which Island any Tortoise may have been brought. When I see these Islands in sight of each other, & possessed of but a scanty stock of animals, tenanted by these birds, but slightly differing in structure & filling the same place in Nature, I must suspect they are only varieties. The only fact of a similar kind, of which I am aware, is the constant

The passage from the ornithological notes where Darwin writes of the different mockingbirds and tortoises (plus the Falkland foxes) replacing one another on adjacent islands: his earliest written suggestion of evolution.

174

Ornithology — Galapagos

reputed difference between the Wolf-like Fox of East & West Falkland Isds. — If there is the slightest foundation for these remarks the zoology of Archipelagoes — will be well worth examining; for such facts would undermine the stability of Species

3308 Yellow-breasted Tyrannus: Female: Chatham Isd: Found in damp places
3309 Scarlet Do. Male
3310 Wren Female
3312 Fringilla — Male
3313. Do. (Sex unknown)
3314 Do. Female
3315 Do Do
3316 Do. Male
3317 Do Male
3318 Do Male
3319 Do Male

(3314–3319: V. Supra.)

3320, 3321, 3322, 3323: (Icterus 3320: Male, jet black) (3321: 3322. Males) (3323. Female). This is the only bird, out of the number which compose the large irregular flocks, which can be distinguished from its habits. — Its most frequent resort is hopping & climbing about the great Cacti, to feed with its sharp beak, on the fruit & flowers. — Commonly however it alights on the ground, & with the Fringilla in the same manner, seeks for seeds. The rarity of the jet black specimens is well exemplified in this case; out of the many brown ones which I daily saw, ~~thus~~ I never could observe a single black one, besides the one preserved. Mr Bynoe however has another specimen; Fuller in vain tried to procure one. — I should add that specimen (3320) was shot when picking together with ~~at~~ a brown one, the fruit of a Cactus.

3324 Fringilla · Male · (young?)
3325 Do — Female. —

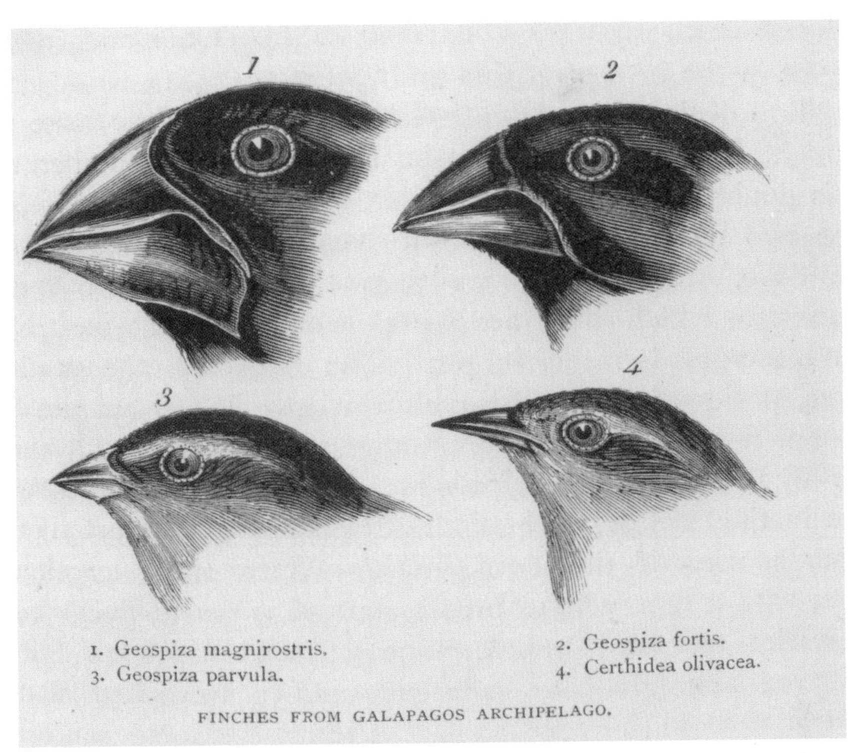

John Gould's comparative drawings of the heads of four species of Galapagos finches. Darwin included this illustration in the second edition of *Voyage of the Beagle*, published in 1845. In the accompanying passage, Darwin says that "Seeing this gradation and diversity of structure in one small, intimately related, group of birds, one might really fancy that from an original paucity of birds in this archipelago, one species had been taken and modified for different ends." This was the earliest published hint at his already well-established evolutionary ideas.

134

Co. p. 461 — Lower Silurian — several existing genera: Nautilus, turbo. buccinum. turritella. Terebratula, orbicula. with many extinct forms & Trilobites

Sept 25th. In considering infertility of hybrids inter se. the first cross generally better & sisters & therefore somewhat unfavourable. —

Is not [illegible] cases in title he thinks disease has opposed the increase of animals nearly proportion to the numbers that can live. —

28th. We ought to be far from wondering of changes in number of species, from small changes in nature of locality. Even the energetic language of ~~Malthus~~ Decandolle does not

increase of better, must be prevented solely by positive checks, excepting that famine may stop desire. —

convey the warring of the species as inference from Malthus — in nature production does not increase, whilst no checks prevail, but the positive check of famine & consequently death.

population in increase at geometrical ratio in far shorter time than 25 years — yet until the one sentence of

there is spring, like food used for other purposes as wheat for making brandy. —

Malthus no one clearly perceived the great check

Even a few years plenty, makes population in man increase, & an ordinary crop causes a dearth.

amongst men. — take Europe on an average every species must have same number killed year with year by hawks, by cold &c — even one species of hawk

The final cause of all this wedgings, must be to sort out proper structure & adapt it to change.

decreasing in number must affect instantaneously all the

to do that for form, which Malthus shows, is the final effect (by means however of volition) of the populousness on the energy of man

rest. — One may say there is a force like a hundred thousand wedges trying force ~~into~~ every kind of adapted structure into the gaps of in the economy of nature, or rather forming gaps by thrusting out weaker ones.

The famous passage in Notebook D where Darwin begins to characterize natural selection.

A page of Darwin's 1842 *Sketch*. Though his Transmutation Notebooks D and E show that he had discovered the process of natural selection by the late 1830s, this *Sketch* seems to record the first appearance of the actual term "natural selection."

of the results of evolution—the patterns that led him to conclude that life has evolved in the first place, now relegated to the supporting role of confirmed predictions of what the fruits of the evolutionary process should look like. It is fascinating, in this regard, that he pinpoints the pattern he thinks was chiefly responsible for leading Lamarck to conclude that "species are mutable." For not until he published the *Origin* in 1859 would he openly admit that patterns of distribution of "allied" forms in space and time were what brought him to the same basic conclusion as Lamarck.

Isolation remains important in Darwin's thinking in 1844. It is telling, though, that his discussion of the importance of isolation in the generation of new species is for the greatest part confined to the second section of his essay—in a section entitled "An attempt to explain the foregoing laws of geographical distribution, on the theory of allied species having a common descent"—rather than the first section on the mechanisms underlying the production of variation and on selection. Darwin's organization of subject matter has had an enormous subsequent effect on the conduct of evolutionary theory—isolation being a perfect example. Some people (myself among them) agree with the nineteenth-century biologist G. J. Romanes when he said, "Without isolation, or the prevention of free intercrossing, evolution is in no case possible"; others have spent their entire careers in evolutionary biology hardly mentioning the nature and significance of isolation at all.

But to me by far the most arresting and telling section of the 1844 *Essay* is Darwin's handling of the fossil record, especially the simple fact that, though larger-scale changes through time agree with his expectations, the actual recorded histories of individual species lineages do not and must be explained away by blaming the vagaries of the fossil record itself. But more than this, his by now familiar approach, is Darwin's closest flirtation with the patterns that Cuvier had seen so long before—patterns in the history of life that only now are beginning to attract serious attention from evolutionary biologists:

> I need hardly observe that the slow and gradual appearance of new forms follows from our theory, for to form a new species, an old one must not only be plastic in its organization, becoming so probably from changes in the conditions of its existence, but a place in the natural economy of the district must come to exist, for the selection of some new modification of its structure, better fitted to the surrounding conditions than are the other individuals of the same or other species. (pp. 163–64)

No Cuvier here; just a beautifully concise statement of his views. But Francis Darwin, in his 1909 edition of this essay (reprinted verbatim by De Beer), footnotes the end of this passage—giving Darwin's annotation, and making a comment of his own:

> [Note in original] Better begin with this. If species really, after catastrophes, created in showers over world, my theory false.
>
> [Francis Darwin then adds] In the above passage the author is obviously close to his theory of divergence.

Previously (as well as later, in the *Origin*), Darwin mentions groups of related species possibly arising and disappearing in bursts; but here, in his own annotation, he is referring to the even more general pattern of extinctions and subsequent evolutionary bursts of biotas—the sorts of patterns Cuvier had in mind in an otherwise nonevolutionary context, and the very same patterns which many of us now find crucial to understanding how the evolutionary process actually works. "My theory false" means only his vision of gradual evolution of species under natural selection as set forth in the passage quoted immediately above. But neither the entire notion of evolution, nor the process of natural selection, is falsified by admitting that species are "really, after catastrophes, created in showers over world." Rather, admitting the existence of such repeated patterns in life's history merely means that we need to rethink the environmental context in which natural selection works.

Francis Darwin is also right that Darwin's phrase "a place in the natural economy of the district must come to exist" foreshadows the only later addition to his theory to come after 1844: the "theory of divergence," which essentially says that new species, and hence new adaptations, or "evolutionary novelties," will survive only if they afford new means of making a living in what Darwin so aptly and presciently refers to as the "natural economy." That the probability of new species surviving has everything to do with their occupying a niche somewhat different from all other existing species (especially the parental species should they come into contact) is a theme that is only recently coming back into mainstream evolutionary thinking. Indeed, Darwin comes close to seeing this when he remarks that there seems to be a correlation between the geographic extent of a fossil species and its persistence in geological strata. In a very important comment that, had he pursued it, would have eliminated his concerns that the fossil record did not agree with his theory, Darwin wrote: "As we see some species at present adapted to a wide range of conditions, so we may suppose that such species would survive unchanged and unexterminated for a long time" (p. 165). He goes on to say that how some species come to be broadly adapted compared to others "is of difficult explanation." In this he is right—but the theme has come back in evolutionary theory in the latter part of the twentieth century, playing an important role in recent considerations of rates of speciation and extinction.

On that score, there are several other themes in this 1844 *Essay* that have a strikingly modern feel to them—ignored as they have been for the most part of the intervening years to the present moment. For example, Darwin says that crossbreeding, and not selection, is responsible for the uniformity of species, in addition to his thoughts on the differential breadth of species' adaptations, a harbinger of what has emerged (as we shall see in chapter 5) as the main explanation for why species remain so uniform through time as well as over space. He also clearly enunciates (p. 114) what many years later came to be called the "founder effect"—where a few individuals of both sexes (or even just a single pregnant female) might invade a new territory, and under the "new conditions of existence," bringing only a small sample of existing variation in the parental species, undergo rapid evolutionary change under natural selection.

Darwin's thoughts have become pretty much settled and his rhetoric crystallized in this *Essay* of 1844. He ends with the passages on the East Indian rhinos as well as his ringing conclusion about the "grandeur in this view of life." And while there is, in his surviving correspondence, some discussion of various points once he confesses his "murder" to Hooker—also in 1844—there is no further written documentation of his evolutionary thinking until the late 1850s. He began his magnum opus, tellingly entitled *Natural Selection*, in 1856 (R. C. Stauffer published his transliteration of the second part of this manuscript in 1975). Well underway when Wallace's manuscript reached him in June 1858, Darwin scrapped what would have been a tediously overlong treatment in favor of what was, even so, to prove itself a much longer treatment than he had written so far, the 1859 "Abstract": *On the Origin of Species by Means of Natural Selection; Or the Preservation of Favoured Races in the Struggle for Life.*

On the Origin of Species

By the time Darwin brought himself to publish his evolutionary views, his rhetorical path was well trodden, albeit known to none but Emma and himself. True, he had corresponded with a small inner circle of friends and colleagues. But no one had had the opportunity of becoming familiar with the full *Essay* of 1844, for example, so his overall theory and mode of argumentation was known only fragmentarily—and then to but a small number of people.

Publication of the *Origin* in 1859 changed all that, of course. It is a mature work in the best, and worst, senses of the term. It seemed fresh to all those who bought out the first edition—and benefited in that regard simply for having gone through the various iterations, allowing Darwin to hone his logic as he mustered his arguments and piled example upon well-chosen example. Only the benefit of reading the earlier iterations shows the *Origin* for what it truly was: not the work of someone transported by a creative frenzy, but rather the carefully crafted exposition of someone for whom the idea had long since ceased to be novel. The excitement of the early days is dimmed. And if Darwin was able to put his very best argumentative foot forward after years of consideration and practice, he was also more firmly than ever committed to his vision of how natural selection, working on variation (which he persisted in seeing induced in large measure from "external conditions"), brought about the entire evolutionary history of life. To those of us who have had the benefit of reading his earlier, private notes and essays, there is little new to be found in the *Origin*.

Though by now Darwin had dropped the two-part division of process (variation, artificial and natural selection) and pattern-as-expected-outcome of evolution (geography, gross patterns of the fossil record, classification, unity of type, comparative anatomy, and embryology—with special attention lavished on instincts and the disappointing fossil record of species lineages), nonetheless the fourteen chapters of the first edition of the *Origin* follow the familiar sequence of topics of the two earlier essays. By now reaching 490 printed pages, the beefing up once again consists of many more examples, culled from the literature and correspondence—and from Darwin's own experiments and observations made in the interim.

Most notably among the latter is Darwin's firsthand experience with pigeon breeding. He writes: "Believing that it is always best to study some special group, I have, after deliberation, taken up domestic pigeons" (p. 20 of the original edition). Noting (correctly) that all domestic varieties are derived from the ancestral rock dove *Columba livia*, Darwin discusses reversion to "wild type," as well as the generation of the highly distinctive breeds through artificial selection. His experiences with domestic pigeons stand in sharp contrast with the eight years he earlier spent monographing barnacles.

Barnacles are mentioned only briefly in a few passages in the *Origin*, reflecting Darwin's conviction that the patterns of classification are expected outcomes of his theory of evolution through natural selection and not sources of insight into the evolutionary process itself. This attitude was to persist throughout the bulk of the subsequent history of evolutionary biology.

Darwin extends his commitment to blurring the distinction between varieties and species in the *Origin*. Of species and subspecies, he notes (p. 51) that "these differences blend into each other in an insensible series; and a series impresses the mind with the idea of an actual passage." And though reproductive isolation looms in the background, he now moves away from a clear, reproductively based species definition in favor of the view that species are what an experienced naturalist judges them to be. His emphasis here is on seeing species as groups of similar organisms rather than focusing on reproduction solely—the aspect that, after all, ensures simultaneously that species are both relatively uniform in appearance (because of interbreeding) and distinct from closely related species (because species as a rule do not interbreed). However, he continues to stress that all degrees of hybridization between species are encountered in nature.

Thus, well-marked varieties are incipient species—and so forth. Darwin has finally dealt with Whewell's (1837) dictum that "Species have a real existence in nature, and a transition from one to another does not exist" by all but eliminating the "real existence" of species as discrete entities in nature in order to establish that, in point of fact, a "connection" between species (i.e., evolution) does indeed exist. It would be many years before the reality of species was restored in evolutionary biological thinking—and the status of species still remains debated but, in some circles, virtually ignored. Long before he wrote the *Origin* Darwin came to think that the differences between closely related species, and indeed any geographic or stratigraphic data suggesting species stability, was antithetical to his view that evolution proceeds through the modification of adaptive features by natural selection. As biologists such as Ernst Mayr have long remarked, Darwin did not solve the problem of how species originate (though there is plenty in his works dealing with the notion of the derivation of species through geographic isolation) so much as he downplayed the very existence of species whose origins require some special theory to explain.

Isolation still plays a role in Darwin's thinking. But I agree with those historians who see that role diminished over his initial stronger reliance on isolation as necessary for the emergence of new species, and for the prevention of the pervasive tendency for

Charles Darwin in 1854. He was soon to begin work on his "great species book," finally determined to bring his evolutionary theory out in the open.

interbreeding to erase differentiation of populations of a species living in different regions. Indeed, after listing the familiar reasons why isolation must be considered important, he goes on to say that

> Although I do not doubt that isolation is of considerable importance in the production of new species, on the whole I am inclined to believe that largeness of area is of more importance, more especially in the production of species, which will prove capable of enduring for a long period, and of spreading widely. (p. 105)

Thus the rhealike patterns—of closely related species replacing one another geographically in large continental areas—come to seem more important to Darwin than the effects of isolation on archipelagos like the Galapagos. Originally they were of equal interest and significance, although it is fair to say that, in the late 1830s and into the early 1840s, it was most especially the Galapagos patterns that seemed to carry more weight in his developing theory.

One novelty of the *Origin* is its single diagram. Tucked between pages 115 and 116 of the first edition, the diagram (reprinted here on p. 195) accompanies Darwin's announcement of his "principle of divergence"—already in fact familiar at least in embryonic form in the 1844 *Essay*, though Darwin himself considered it as a novel thought that occurred to him in 1854. He likens diversification of closely related species to the division of labor of body parts within a single organism—a somewhat far-fetched analogy. But he makes the undoubtedly valid point (pp. 111ff.) that the more diversified a new species is, the greater its chances for finding its own unique spot in the economy of nature—a theme (as already noted) that only recently has come back into evolutionary thinking.

The remaining themes of the *Origin* are already quite familiar. Darwin uses his single diagram to illustrate his discussions of two of his patterns, classification and comparative anatomy (including embryology). The diagram and its occasional use throughout the *Origin* make it clear that Darwin saw that one of the two main "expectations" of "descent with modification" would be the production of a branching array of ancestors, descendants, and the production of collateral lineages of varying degrees of relatedness ("propinquity of descent"). Hence the nested groups (genera, families) in Linnaean classifications are what one would expect to find given the process of descent with modification, and the classification mirrors the array of progressively similar anatomical features and embryological stages.

The second grand prediction of evolution is that there ought to be a basic sequence of simpler to more advanced forms in the fossil record—and points of resemblance in a gross sort of way between older and progressively younger organisms recovered from the fossil record, a familiar point that Darwin now makes by enlisting the aid once again of his diagram. He reiterates (p. 317) his belief that, although "on the theory of natural selection the extinction of old forms and the production of new and improved forms are intimately connected together," nevertheless, "the old notion of all the inhabitants of the earth having been swept away at successive periods by catastrophes, is very generally given up. . . ." He simply does not consider the slightly modified form of the pattern: that many, but by no means all, species of a region (or even, in rarer cases, over the entire earth) may indeed be "swept away" by extinction, and what the underlying evolutionary dynamics might be. Rather, he insists that for the most part extinction is a gradual process, largely the result of competition between older species and the newly formed, later species that are destined to outcompete their progenitors, literally replacing them by driving them to extinction. Today's thinking has it the other way around: extinction and evolution are indeed causally linked, but it

A page of Darwin's notes, written in 1855, on which he develops his "Principle of Divergence"—the only substantive addition to his theory after the early 1840s.

is extinction that triggers evolution. New species tend to arise and survive (through Darwin's "principle of divergence") only after a significant number of existing species are driven to extinction, almost always through "catastrophic"—or at least overwhelming—changes in the physical conditions of life.

Darwin persists in thinking that the appearance of catastrophes—and the lack of examples of gradual change within species documented by paleontologists in the fossil record—is to be blamed on the nature of formation of the stratigraphic record, on its tendency to be eroded or metamorphosed to oblivion, and on the difficulties of finding, extracting, and studying fossils in the first place. He adds (p. 342) his famous ringing conclusion: "He who rejects these views on the nature of the geological record, will rightly reject my whole theory." He never really considered the possibility that, though he was more right than wrong about the extreme age of the earth and the vagaries of the fossil record, nonetheless the patterns he saw as a young man examining South American fossils—patterns that led him, along with his guanacos and rheas, to posit saltational evolution in his very first jottings on evolution in the Red Notebook—themselves had more truth than not in them. Those patterns were fraught with implications about the evolutionary process. But, overwhelmed with visions of continuity derived from his mental picture of natural selection working on entire species, no matter how widespread, Darwin turned his back on the fossil patterns that had led him to his evolutionary ideas in the first place.

It is worth emphasizing again that Darwin's adoption of the hypothetico-deductive method is thus not absolute. He clings to his theory and throws out the fine-scale fossil evidence as unreliable. He refuses to reject his theory because fossils do not yield insensibly graded species. This is of course just as well—as his theory *was* true. What was inaccurate instead was his predictions of what the fossil record should look like if his theory of evolution by natural selection were correct. He was simply wrong when he imagined that natural selection could transform an entire, far-flung species slowly and gradually through geological time—in exact parallel to the success of cooperative, local breeders in artificially selecting traits in their various breeds.

And that's just about it. Darwin's *Origin* is a mature and generally well-written exposition of his ideas as they stood in the late 1850s, but which were already essentially in place by the time he wrote his two essays in the early 1840s. Indeed, they were largely there when he wrote his notebooks, brimming with creative enthusiasm and inductive insight in the late 1830s as a young man just returned with a head crammed full of impressions and a thirst to join the newly emerging ranks of the men of science.

Darwin's Evolution After the *Origin*

The *Origin* was reprinted several times in what were somewhat misleadingly called successive editions. It was really only the sixth edition of 1872—the version still most commonly read today—that contained significant emendations and additions. None of these represented major changes in Darwin's personal thinking. Rather, they were responses (largely grudging and partial) to criticism. For though Darwin had tried his very best to anticipate all possible objections, and to counter them all, critics refused to back away without an argument. For example, because several paleontologists in their reviews mentioned Darwin's reluctance to admit that species typically remain more or less unchanged throughout often long spans in the fossil record, he did admit that some species apparently remain stable through long periods of time. Similarly with his discussion of the difficulties in positing the gradual evolution of structures as complex as the human eye; yet St. George Mivart nonetheless attacked him on this very point—necessitating still more discussion in the sixth edition.

Darwin is also said to have backed off from his original insistence that absolutely all evolution occurs through natural selection, especially being charged with embracing to an extent greater than before the inheritance of acquired characters—a position most closely associated with Lamarck. This is to some extent true; but it must be remembered that Darwin even in the *Origin* openly embraced the closely related position that variation is induced by external conditions, especially as those conditions change.

So there is no real further "evolution" of his main lines of evolutionary thinking. *The Descent of Man and Selection in Relation to Sex* (1871), which contains a much fuller (and, I think, more explicitly clear-cut) discussion of sexual selection, adds little truly new to his canon of thought. It is striking, though, that in one passage in this, his last book actually devoted to evolution per se, Darwin does revert to a stronger, or at least a less equivocal, position on the importance of reproductive isolation. Once species emerge—when reproductive isolation is attained—they become independent entities, following their own separate evolutionary fates, and eventual extinction. Without reproductive isolation, whatever adaptive changes might have appeared in the differentiation of a species into "races," "subspecies," or "varieties" always stands the chance of resorption—disappearance through crossbreeding, the tendency for species to remain homogenized, counteracting the differentiation that is bound to come through natural selection in different populations in different regions of a species' entire geographic range. Species are "permanent varieties."

More subtle still are the nuances in Darwin's thought as revealed in his voluminous correspondence after the publication of the *Origin*. Isolation, for example, plays a prominent role (as we shall shortly see) in his correspondence with the German biologist Moritz Wagner. But, once again, as far as the world could see, Darwin's evolutionary thinking in the *Origin* (whether first or sixth edition, it really doesn't all that much matter) was what convinced the world that life had evolved. It also pretty much locked into place how the entire subject is to be approached—right on down to the present day.

CHAPTER 5

Evolution After Darwin

Nearly a century and a half has passed since the *Origin* first appeared in print. Advancing technology has spurred tremendous growth in scientific knowledge—with an emphasis, almost inevitably, on probing the hidden depths of the physics and chemistry of the invisibly small, and probing, as well, the ultra-large dimensions of the universe. Biology has grown vastly more complex. Not only do we now know the basic reasons why organisms vary, and why variation is heritable, we are also beginning to get a grip on how the information ensconced in DNA is translated into the development of fertilized eggs into adults.

But the world—despite the rapid loss of species triggered by human population growth and despoliation of the earth's surface—is recognizably much the same as when Darwin considered it. And his fundamental questions remain: questions about the pulse and pace of evolutionary change, and how the various factors he and others have identified as ingredients of the evolutionary process meld together to produce the evolutionary history of life on earth. The story of evolutionary theory after Darwin left the stage is, of course, long and intricate, made more so as fields have fragmented as knowledge has grown. Darwin was almost certainly the last person to have a sufficient grasp of all the relevant biological and geological information to tackle all the issues of evolution. The division of evolutionary biology into such fields as genetics, ecology, systematics, evolutionary developmental biology, paleobiology, and so on has been not only inevitable but "a good thing" for the growth of knowledge. But the fragmentation into fields has also—equally inevitably—engendered turf wars and gross intellectual misunderstanding. Most paleontologists don't know an awful lot about molecular biology, and vice versa for molecular biologists.

The task here is not to review absolutely everything that's been said in evolutionary biology since Darwin's day, nor to review or even evaluate the ongoing battles in evolutionary biology. Rather, we should look at Darwin's issues afresh, pretending we are Darwin and asking what people now think about the genesis of heritable variation; natural selection; the role of isolation in the formation of new species (and what species *are* for that matter); the relation between extinction and the appearance of new species; and what, if anything, geologists and paleontologists make of the old gradualism versus catastrophism debate these days. We can think of evolution as the fate of heritable information—and take a quick journey through the last 150 years or so focusing on these specific issues.

Darwinian Fallout

Reviews of the first edition of the *Origin* were—naturally—mixed. Most of the reviewers thought Darwin had something to say of potential importance to solving the "mystery of mysteries," but tempered their admiration with the usual "more work must be done" sort of prevarication. Though Darwin's expertise was generally respected, and his exhaustive research admired, few thought the problem of the origin of species was solved once and for all.

As for the men of science whose good opinions Darwin most craved, few were converted. Some, like Sedgwick and Henslow, remained adamantly opposed on largely religious grounds—Sedgwick especially acerbically so. Owen was transparently jealous; but it must also be said that, as a rule, the comparative anatomists of the day were wedded to notions of "archetypes," and so were far more inclined to see stability in anatomical systems than to think it possible that complex structures in one organism's body could be transformed into radically different structures in another type of organism's body.

The exception was Thomas Henry Huxley, Darwin's "bulldog," who took the fight to scientific meetings and other public venues, including of course the most famous of all, the debate with Bishop "Soapy Sam" Wilberforce at Oxford on June 30, 1860. Huxley was an anatomist, as well, and a young rival of Richard Owen's. The degree to which Huxley's enthusiasm for evolution was based on his rivalry with Owen is impossible to judge. Huxley was of course a brilliant scientist, capable of making up his own mind on scientific issues independent of his political rivalries in the profession. But it is worth noting again that, as a comparative anatomist, Huxley never fully embraced Darwin's strict insistence on gradual continuity in the evolution of life. That was a morphologist speaking. Evolution occurs, Huxley agrees, but the transitions between different anatomical designs can at least sometimes be rapid, all-at-once, reflecting their inherent stability.

But the reactions of most of the known figures in Darwin's life—men who were his mentors, contemporaries, and, later, his junior colleagues—must have been eminently predictable. And gratifying as it must have been to Charles that Lyell eventually came around, if not to become excited about Darwin's evolution, at least to admit that he was likely right, it must have come as a bitter disappointment that neither William Whewell nor, saddest of all, John Herschel accepted his theory. Whewell evidently could not stand change in scientific views; and perhaps, after all, Darwin could not have rationally expected Whewell to abandon his religiously tinged views of the nature of things in favor of evolution.

But John Herschel? This was the man whose "mystery of mysteries" inspired Darwin to make his mark on the world by solving the problem of why new species tend to replace those that are lost to extinction. Herschel had articulated the holy grail of natural science—the formulation of a naturalistic explanation of his mystery of mysteries. How bitterly disappointing it must have been to Darwin to have his core idea of natural selection dismissed as the "law of higgledy-piggledy" by the very man who had encouraged him and given him strength in his pursuit!

But no one should ever read his own reviews—good or bad. Darwin claimed he learned much from some of the more pointedly critical comments, but elsewhere wrote that he had gotten quite sick and tired of reading bad reviews. What matters is the work that was done in evolutionary biology as a consequence of his having basically succeeded in convincing the thinking world—especially its younger scientific practitioners—that life has evolved.

Comparative Anatomy and Systematics. Most of the work done in the name of evolution in the decades immediately following the publication of the *Origin* consisted, in effect, of translations of the old way of looking at things into the new evolutionary paradigm. Treatises were written (as they continue to be to the present day) on the evolution of this group or the other. Huxley came to America and viewed the impressive series of horse fossils excavated and studied by the paleontologist Othniel C. Marsh at Yale University, writing back to Darwin about the wonderful sequence beginning with small animals with many toes on up to the emergence of the large, single-toed horses of the present day.

In Europe, the German zoologist Ernst Haeckel transformed the concept of "archetype" into the notion of the common ancestral body plan of entire groups. Systematists immediately embraced Darwin's point that the classification scheme of Linnaeus works so well because it does indeed reflect the underlying structure of the natural world—not as a reflection of God's creative plan, but rather as a simple consequence of the evolutionary process. Organisms group readily into natural groups because evolution—"descent with modification"—means that as new species evolve from old, lineages are created of varying degrees of relatedness. Various species of squirrels belong to the genus *Sciurus* because of the recency of their common ancestry; ultimately, all life—from bacteria to redwood trees—is joined up, sharing features derived from the earliest forms of life on earth, now known to be the macromolecules of heredity DNA and RNA.

But few if any of these enthusiastic converts to evolution had anything substantive to say about *how the evolutionary process works.* Yes, Huxley made that comment to Darwin about saltation; but the initial burst of work under the evolutionary rubric was mostly confined to documenting examples of various aspects of the results of evolution. There was little serious research into evolutionary mechanisms per se—with the exceptions noted below.

And that, I think, is a situation that arose from the instant legacy created by Darwin's presentation of his ideas in the *Origin*, a handcuffing of evolutionary thinking that in my opinion persists right down to the present day. Darwin was almost too clever. He followed his own suggestion, jotted down on page 118 of Notebook E, to rethink all the clues, the patterns that led him to evolution in the first place—to see if they would flow naturally as predicted outcomes of the process of natural selection over space and through prodigious amounts of geological time. As we have seen, in developing the notion that there is a gradation from varieties to true species over space—and by claiming that the lack of smooth, gradual continuous change in the fossil record is the artifact of a poor record, and not evidence against such smoothly gradual, continuous change—Darwin effectively rewrote the smaller-scale patterns of

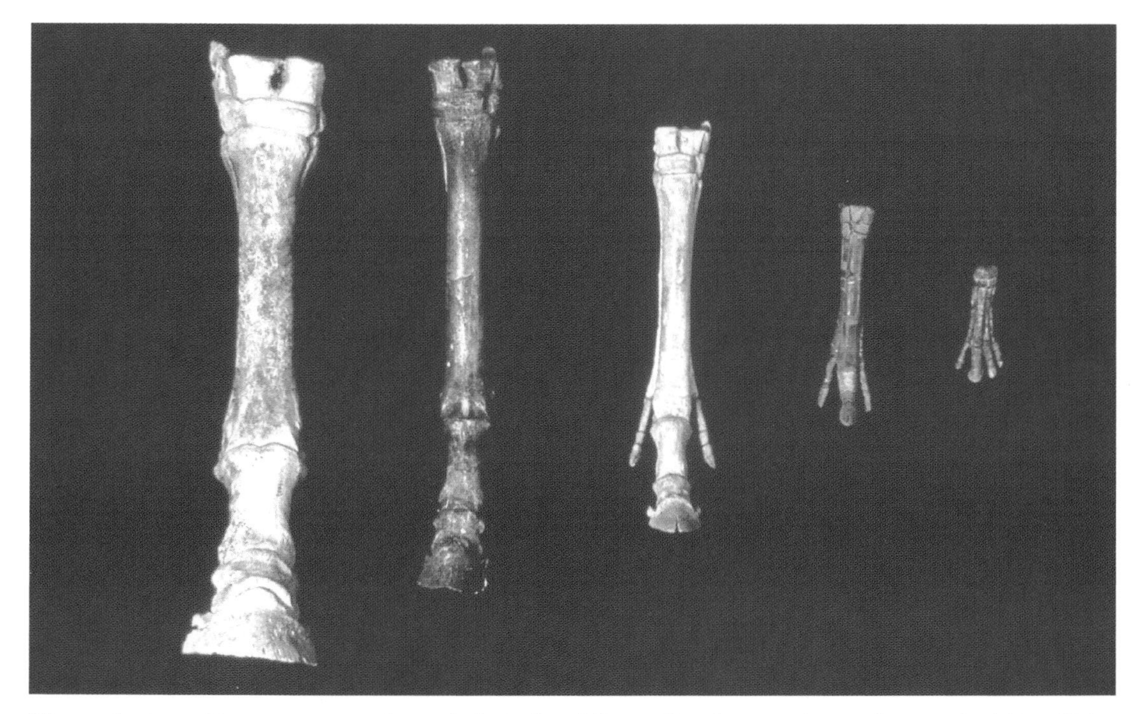

The evolution of horses as seen through their feet. The earliest horses (*Hyracotherium*) had four digits (toes) on the front legs and three on the rear. Modern horses (*Equus*) have just one digit, front and rear. The fossil record reveals a complex array of horses through time, with reduction in toe number happening progressively in stages over the past 50 million years. Horses provide a very clear example of significant evolutionary change preserved in the fossil record.

change within and among species into predicted outcomes of his vision of natural selection. All of the other patterns are actually predicted consequences of the simple idea of evolution; not, as Darwin so often claimed, of any special theory of how evolution works, or the manner and degree to which natural selection is the central element of the evolutionary process. These included large-scale biogeographic distributional patterns; the large-scale patterns of the fossil record; the nested sets of resemblance linking up the anatomical organizations of large groups of animals and plants (and, by extension, all of life—Darwin would have loved to have known about DNA and the many other exciting discoveries linking up the largest groups of life on the planet that have been discovered since his time); the consequent progressive similarity in groups of embryos; and the reflection of all this into large-scale biological classifications.

In so doing, Darwin masked from his successors the true nature of his own thinking when he let nature come to him and discovered the grand truth of evolution. In the early days, before the natural patterns became predicted outcomes of evolution/natural selection, such patterns were fraught with implication: not just that life has evolved, but also with suggestions about what the evolutionary process itself must consist of. As soon as they became secondary lines of evidence—tests of the idea of evolution and of the efficacy of selection—they seemed less interesting as lines of investigation into the nature of the evolutionary process itself.

This is why Darwin hardly mentions barnacles in the *Origin*, even though he had spent eight years on an exhaustive study of barnacle systematics. It was many a year before a systematist was to write anything cogent about the nature of the evolutionary process, and systematics remains to this day a vastly underutilized source of inspiration for the generation and testing of ideas on the evolutionary process. Most works on systematics hardly bother to acknowledge that the patterns of relationship uncovered in the analytic work reflect the evolutionary history of the group (this is taken for granted), let alone venture to "speculate" (a word Darwin liked) on what the implications of the distributions in space and time of the species linked into a particular network of evolutionary relationships might have for our larger understanding of the nature of the evolutionary process. Our grasp of the process is the poorer for this.

Paleontology. The case of paleontology has been even more difficult, marginalized as it has been through the far-reaching and lasting success of Darwin's argument that the fossil record is too poor in detail to trust it to tell us anything whatever about the nature of the evolution of species. Nor is it just non-paleontologists who have been prone to swallow Darwin's dismissal of the (apparent) disagreement between the predicted outcome of his theory and what the fossil record actually shows—as the expected outcome of the inherently spotty preservation of fossils in the first place. Paleontologists for the most part have bought the line as well. The dearth of examples of slow, steady evolutionary change within fossil species, supposedly attributable solely to a uniformly spotty fossil record, has become what Stephen Jay Gould called "paleontology's trade secret."

Indeed, in my opinion it was not until George Gaylord Simpson, an outstanding American vertebrate paleontologist of the mid-twentieth century, became caught up in the heady days of evolutionary revival in the 1930s and 1940s that the promise of paleontology as a source of insight into the nature of the evolutionary process was rediscovered. Simpson thought that, whatever use one makes of what he calls "evolutionary determinants"—nature and rate of mutation, population size and other variables of genetics—to concoct a theory of evolutionary mechanisms, that theory must be tested against actual patterns in the fossil record. To Simpson, what happens to one hundred laboratory rats over ten years might be interesting, but not necessarily an accurate guide to what happens to a billion rats over 10 million years.

In the Preface to his first and arguably most creatively original book on evolution, *Tempo and Mode in Evolution* (1944), Simpson summarized the situation in evolutionary biology trenchantly, though not without humor:

> Not long ago paleontologists felt that a geneticist was a person who shut himself in a room, pulled down the shades, watched small flies disporting themselves in bottles, and thought that he was studying nature. A pursuit so removed from the realities of life, they said, had no significance for the true biologist. On the other hand, the geneticists said that paleontology had no further contributions to make to biology, that its only point had been the completed demonstration of the truth of evolution, and that it was a subject too purely descriptive to merit the name "science." The paleontologist, they believed, is like a man who undertakes to study the principles of the internal combustion engine by standing on a street corner and watching the motor cars whiz by. (p. xv)

There was no genetics in Darwin's day, but otherwise Simpson perfectly caught the dichotomy Darwin set up. The mechanisms of evolution could only be addressed through the principles of generation and selection of heritable variation; all other patterns of the natural world were to be seen as predicted outcomes of evolution. Hence, as Simpson says, validations of the overall fact of evolution (though Darwin would have added "through natural selection"). Where they didn't validate the idea—namely, the fossil record of species-level evolution—we throw out the evidence as imperfect, rather than rethink our ideas. Simpson was saying that we can do better: we can test our ideas of the evolutionary process—specifically of the generation and selection of heritable variation, about which much was already known by the 1930s—*and correct them if they do not accurately predict what we observe in nature.*

Coming from a paleontologist, these were bold words. Simpson himself stuck to the old Darwinian line that the lack of good examples of fine-scale, species-level gradual change in the fossil record was indeed the artifact of a lousy record. But he also said that larger-scale patterns of abrupt appearance of major groups of organisms (horses that graze grass evolving from leaf-browsing ancestors) reflected not a poor record that fails to preserve millions of years of insensibly graded change, but rather the fact that evolution of such novel groups typically happens very rapidly, in quick bursts of evolutionary change he dubbed "quantum evolution." That example—Simpson maintained that the fossil record has real meaning, true "signal," and not just "noise"—was to prove inspirational to many paleontologists who have come along after him, trying to wrest evolutionary meaning from the fossil record of the history of life.

Some nineteenth- and early twentieth-century paleontologists after Darwin did give evolution a good try, though. Convinced that dense sampling of well-preserved specimens through thick sequences of strata would reveal the "insensibly graded series" that Darwin thought must be there if the fossil record was sufficiently completely formed, preserved, and sampled, they made valiant efforts to square their science with Darwinian evolution. The most famous of such studies was that of A. W. Rowe, who published a paper in 1899 on the evolution of heart-urchins of the genus *Micraster* from the chalk deposits of the English Cretaceous Period. Rowe claimed to find gradual change linking up a series of species through time. What he actually had was a series of closely related species—some of which were possibly related in an ancestral-descendant fashion—that nonetheless do *not* show the long periods of gradual change taking one species into another through the passage of geological time. Just like Marsh's horses that Huxley saw, Rowe and other paleontologists had no trouble collecting sequences of closely related species through time. But at the microlevel, the gradual transition from one species to the next that Darwin demanded on first principles from his concept of natural selection was really not there in any of these early studies purporting to find Darwinian gradual evolution in fossil series.

Embryology. There is one field of study that, at a low level of intensity but a more or less constant rate, has produced creative scholars who have contributed to ideas on the evolutionary process beyond simply demonstrating that their patterns are consistent with the idea of evolution. That field is embryology, the comparative anatomy of embryonic development, with its general pattern of greater overall resemblances between embryos at earlier, rather than later, stages of development—the very pattern that Darwin in his *Autobiography* tells us he was most proud of developing as one of the predicted patterns of the evolutionary process.

Not all embryologists (developmental biologists) are drawn to evolution, of course, for there has been much to analyze in development for its own sake. A worthy successor to Herschel's "mystery of mysteries," the process that takes a fertilized egg through all developmental stages up to a sexually mature adult remains something not yet fully understood, even though the progress, especially in the post-molecular world of the late twentieth and early twenty-first centuries, has been extraordinary.

Yet, probably because embryonic development is itself a process, and the emergence of fully formed adults can be observed from start to finish, developmental biologists from the very beginning have not been as reluctant as systematists, pale- ontologists, biogeographers, and other "pattern" biologists to join actively in the ranks of those interested in evolutionary mechanisms. Karl von Baer, a contemporary of Darwin's and thus not, in a strict sense, an evolutionist, nonetheless formulated general "laws" of embryonic development that served as the basis of the work of later embryologists such as the enthusiastic Darwinian Ernst Haeckel, and later W. Garstang and especially Gavin De Beer, whose publication of Darwin's early manuscripts helped keep the important early scholarship of Francis Darwin alive.

Today's "evolutionary developmental biology" is coming very close to specifying how changes in timing of the regulatory portion of an organism's genes can lead to the emergence of differences between closely related species. The 98.6 percent genetic similarity between chimps and humans, for example, is now routinely interpreted as the relative effect of developmental timing rather than concluding that it is the 1.4 percent of the genes that humans and chimpanzees do *not* share that makes all the difference between us and the chimps.

The developmental biologists of earlier generations were of course not privy to the methods and results of molecular genetics, yet they were still able to pinpoint just where in embryological development changes would appear that would result in permanent evolutionary change. If new features were simply "tacked on" to the end of a developmental sequence, it would look as if an embryo passed through all the embryonic stages of its ancestor, the differences only emerging at the end—the source of Haeckel's famous dictum, "ontogeny recapitulates phylogeny." In contrast, Gavin De Beer many years later championed the opposite sort of effect: if embryonic development began to stop short of the final stages of the ancestor, the descendant species would look more like the young of the ancestor. The opposite of Haeckel's "recapitulation," De Beer's "neoteny" seems to fit human evolution well: adult humans more closely resemble young than fully adult chimpanzees.

Thus embryology, at least in a muted sense, is the exception to the rule that the structure of Darwin's argument in the *Origin* stifled process thinking in the various fields of "pattern" evolutionary biology.

Ecology and Micro-Biogeography: The Importance of Isolation. Another theme, dampened by the ambivalent position Darwin ended up with by the time he wrote the *Origin*, is the relative importance of isolation in the evolutionary process. Darwin thought isolation essential in his earlier notebooks and essays; yet he had come to think that divergence without physical isolating barriers was the norm on continents, and hence the role of isolation as seen in island groups like the Galapagos was of lesser significance in the evolutionary scheme of things. But he never did completely abandon isolation as an important evolutionary factor: recall that, in a brief passage in *The Descent of Man*, he saw the reproductive isolation of species as crucial to conserving evolutionary novelties—novelties that might well be lost if they arose in varieties that retained reproductive connections within a single species. And isolation is also related to his principle of divergence: the idea that the more adaptively different (or "ecologically differentiated") descendant species are, the more chance they have of securing their own place in the economy of nature, hence the greater their chances of survival. Darwin's principle of divergence remains underexplored in modern evolutionary thinking.

Thus isolation enters into any discussion both of the genesis and of the preservation of adaptive evolutionary change. It is a part—vital to many theorists, from the early Darwin down to the present day—of the context of evolution: how, when, and where natural selection acts to effect change; and when it might be expected to be conservative, stabilizing species against further change. The role of isolation has never been wholly lost sight of in the days since Darwin, but it was not a dominant part of evolutionary theory until its resurrection in the 1930s.

Nonetheless, a number of nineteenth- and early twentieth-century pattern biologists did take isolation very seriously. In addition to G. J. Romanes, David Starr Jordan (an ichthyologist who served as president of Stanford University) was another biologist dedicated to the importance of isolation in the evolutionary process. But the most important of all those in the early post-*Origin* days who championed the significance of isolation was the German biologist Moritz Wagner, who read the *Origin*, was convinced of evolution, and entered an initially cordial correspondence with Darwin.

The historian Frank Sulloway's essay on Darwin's thinking on isolation contains a

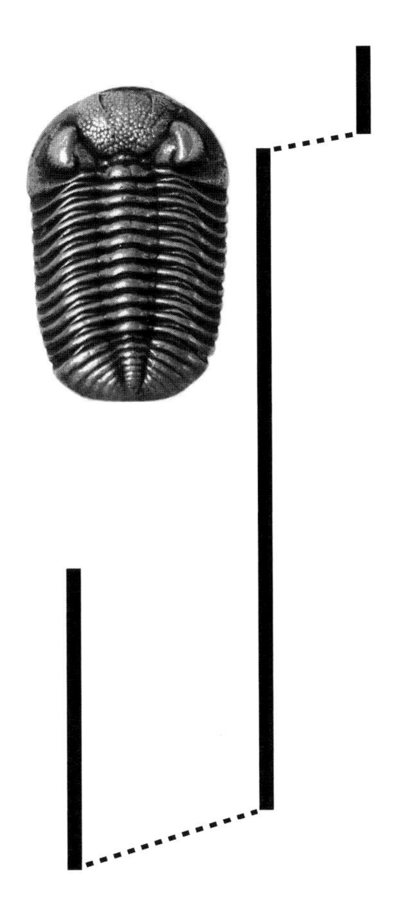

What the fossil record actually looks like: long periods with little or no cumulative change within species, interrupted by relatively brief intervals of evolutionary change, usually correlated with the evolution of one (sometimes more) descendant species from an ancestral species that may or may not survive. The particular example is the evolution of the Devonian trilobite *Phacops rana* complex of species in North America—one of the original empirical examples used in the development of the theory of "punctuated equilibria." The vertical lines represent the temporal duration of individual species that persist relatively unchanged; the dotted lines represent the rapid process of evolution of descendant species from the ancestral species.

fascinating glimpse of Wagner.* There is no question that Wagner's influence on the history of evolutionary thinking has been obscured. His message was seemingly long-ago forgotten—to be resurrected in a general sort of way by later evolutionary biologists (especially Dobzhansky and Mayr in the 1930s and 1940s). But I felt, in reading Sulloway's description of Wagner's ideas, much the same way as I did when I read Darwin's first thoughts on evolution in the Red Notebook where he says evolution happens, and it happens *per saltum*. When I read those words, I thought that, had Darwin stuck to that line of thought, he would not have rejected the fossil record—and my own career as an evolutionary paleontologist simply would not have happened.

That's the way I felt when I later read what Sulloway had to say about Moritz Wag-

*See F. J. Sulloway, "Geographic Isolation in Darwin's Thinking," *Studies in the History of Biology*, 3 (1979).

ner. I had long known about Wagner and his insistence on the importance of isolation in evolution. But I was totally unprepared to find that Wagner essentially derived the components of "punctuated equilibria" (the theory I developed with Stephen Jay Gould in the early 1970s) not, as we had done, from a consideration of patterns of stasis in the fossil record of species lineages, coupled with an appreciation of the theory of geographic speciation, but rather on first principles from his understanding of natural selection, reproduction, and the importance of isolation. Wagner (according to Sulloway) thought that natural selection works to effect change when relatively small populations become isolated, but that, as a new species spreads out and its numbers grow, interbreeding within it will tend to stabilize the species, buffering it against further change.

The pattern Wagner derives as typical is one of rapid evolutionary change as new species emerge from populations of an ancestral species that have become isolated from the rest of the parental species; subsequently there will be long periods of relative stability and little or no further evolution. Darwin thought it was rubbish yet it is indeed what the fossil record shows—and, in a sense (though he denied it), Darwin knew it was true!

Wagner was right about the basic patterns. But we now know that the sort of gene flow that preoccupied Darwin, Wagner, and so many other Victorian-era evolutionary biologists is not in fact the agent that stabilizes—or even homogenizes—species spread out over large geographic ranges. As we will see, there are other interactive effects between geography and gene pools that stabilize species into the long periods of stasis we so typically see in the fossil record.

Genetics. By far the greatest difference between Darwin's world and our own so far as evolutionary biology is concerned is the profound understanding of how it is that organisms resemble their parents, and how heritable variation is generated and maintained in populations. Darwin would be transfixed with joy were he able to come back and catch up on all that we have learned.

Growth in understanding the principles of heredity has come in successive waves—each one with profound implications for evolutionary theory. The first Genetics Revolution, starting in 1900, was so heady, with discoveries coming so fast and furiously, that many biologists thought initially that the newly discovered facts completely superseded Darwin's theory. It took several decades for a full reconciliation of genetics and evolutionary theory to take shape. The second Genetics Revolution, beginning with the Watson, Crick, and Franklin elucidation of the structure of DNA in the early

1950s, has of course provided a deeper, expanded, and rather different picture of the mechanisms of heredity. These mechanisms are now understood on the much more sophisticated level of the molecules themselves: what they are composed of, and how they work in terms of both transmission to the next generation, how they come to be altered ("mutations"), and how genetic information is translated into the production of proteins and, indeed, serves as a template and regulates the timing of the development of organisms from a fertilized egg. The evolutionary ramifications of this second revolution are still being developed.

We now know that some of Darwin's speculations on heredity were wrong. Though it was in fact irrelevant to his theory of evolution by natural selection per se, he embraced a general theory of "pangenesis"—supposing that all parts of an organism's body somehow contribute heritable information to the sex cells (eggs and sperm in animals). That he could be so wrong on the fundamentals of inheritance, yet essentially so right on how selection acts on heritable variation, is graphic evidence that selection theory simply requires that there *be* heritable variation in the first place. How that variation is in fact generated and transmitted to succeeding generations is of great interest and importance—but it is not necessary to know simply to articulate the principle of natural selection.

Yet it turns out that Darwin was also wrong on some issues that were in fact central to his theory. Most significant was his belief that the external environment could stimulate the production of heritable variation. Especially when climatic and other environmental conditions change, Darwin thought that the new variation that selection could act on to adapt an organism to new conditions would somehow be induced. The orthodoxy that was soon to emerge before his century was over steadfastly maintained the opposite: the external environment, while it may induce mutations (e.g., via radiation), does not induce heritable variation that may prove useful to an organism adapting to its changed surroundings. (That there may indeed be circumstances, however, where environmental change effects heritable changes in developmental timing is a radical idea only just now, in the early twenty-first century, being seriously considered; in a sense, Darwin may turn out to have been right on this issue, as we will see later in this chapter.) Darwin also, in later editions of the *Origin*, admitted a related proposition most closely associated with Lamarck: the inheritance of characters acquired through use and disuse.

The man who blew all these suppositions out of the water—including Darwin's theory of pangenesis—was the German biologist August Weismann. Writing primarily in 1870–90, Weismann drew an absolutely critical distinction between the soma ("body") and germ line (eggs and sperm of animals). Weismann proclaimed that only the genetic material—whatever it was—confined to the nucleus of germ-line cells contained heritable information that could be passed from parents to offspring. What happens to the soma—the non-germ-line organs, tissues, and cells—is irrelevant to the characteristics passed along to offspring.

Thus the germ line causes the soma of embryos to develop, and parental characteristics to be passed along to offspring, albeit in what still seemed, in the late nineteenth century, non-straightforward, not wholly understandable ways. Weismann's distinction between the soma and the germ line was the great conceptual advance that enabled the significance of Gregor Mendel's rules of inheritance (once they were rediscovered independently three times right around 1900) to be quickly grasped. From then on the science of genetics was off and running.

It is worth noting, however, that Darwin himself did in a very real sense perceive the difference between the somatic, non-reproductive side of organismic anatomy and behavior and the purely reproductive side of things when he articulated his notion of sexual selection in contrast with natural selection. Sexual selection pertains only to the relative reproductive success of some members of a local population within a species over others, purely because they are better at finding a mate and making babies. Natural selection has to do with the other side of the ledger—who is more successful at survival: finding energy resources (food if you are an animal), and otherwise avoiding death from disease, predation, harsh environmental conditions, and so on. Weismann's distinction between germ line and soma brings Darwin's distinction down to the level of the cell (and much lower, given what is now known about chromosomes, DNA, codons, and so forth). The distinction also travels up the scale to larger-scale biological systems. On the one hand, to ecosystems, where all is held together by the transfer of matter and energy between organisms of different species. And on the other hand to species and higher taxa (genera, etc.), all of which are packages of genetic information. Weismann's distinction clarifies Darwin's distinction between sexual and natural selection; and it goes a whole lot further, letting us see the essential differences between economic and reproductive systems in biological systems of all scales.

The rumor persists that Darwin had on his shelves at Down House an uncut copy of a paper sent by the Austrian monk Gregor Mendel. The supposition is that had

Darwin bothered to cut the pages and read the paper, he would have changed his views radically on the nature of inheritance—with all sorts of implications for his evolutionary theory. There is no truth to the rumor, at least insofar as I have been able to determine.

But it is quite true that the rediscovery of Mendel's work—first published in 1866, but largely ignored until the years right around 1900—caused a radical change in evolutionary thinking. Darwin himself was in relative eclipse in biological circles by 1909—simultaneously the hundredth anniversary of his birth and the fifty-year anniversary of the publication of the *Origin*. Francis Darwin used the twin anniversary to publish his invaluable editions of the 1842 *Sketch* and 1844 *Essay*, and a big celebration was held at Cambridge University. But the mood in biology had changed: from the field to the laboratory, from observations and informal experimental manipulations in nature to what was nearly universally seen as the more precise, exacting work of bench laboratory science. Electricity was now in the lab, and I well remember poring over a massive physiology text published (ironically enough) in 1909 where the author gleefully proclaimed the new status of biology as a full-fledged science, free at last from such old fuddy-duddy natural history notions as Darwin's natural selection.

Biologists on both sides of the Atlantic in the first decade of the twentieth century had much to be excited about. If Weismann pinpointed the locus of material of heredity in the nucleus, biologists soon discovered and named chromosomes—and saw the banding on the giant chromosomes of the salivary glands of fruit flies (the very organisms that "disport themselves in bottles," in George Simpson's witty phrase), calling them genes, as the "particles" of heredity.

Mendel had shown that the inheritance of characters occurred by assuming the existence of such particles ("particulate inheritance"), and that often the particles had alternate, variant forms. Sometimes combining the particles yielded a blend (red and white flowers might be pink if there was one of each particle inherited by the offspring); but often an alternative form of a gene ("allele") would dominate the other—so that the gene producing tall pea plants, for example, would always produce tall plants as long as either one or two of the tall alleles was present; only if the short allele was present in two copies would the plant be short.

Darwin's notion of natural selection assumed predominantly continuous variation; height and length of horses, for example, varies within limits—as does volume of milk production in cows, length of hair on sheep, and so forth. But the early results in genetics tended to support Gregor Mendel's observations of particulate inheritance, with largely either/or effects on the characteristics of organisms. It was hard to square Darwin's views on variation—and how selection could act on it—with the new discoveries in genetics.

Early geneticists also observed the origin of new variants, which they called "mutations." And most of the mutations they found at first seemed to have large-scale effects—very often harmful, or even downright lethal, to the organism that had them. This, too, was out of sync with Darwinian expectations, where new variations would be expected to be, if not immediately useful to the organism, at least contributing to the normal variation of healthy individuals in a population.

As is so often the case when breakthroughs come fast and furious—particularly when much of the new discovery appears to conflict with older suppositions on how things are—there is a tendency to throw out the old in favor of the new. And some of the early geneticists who actually did turn their attention to the evolutionary implications of their results did just that: the Dutch botanist Hugo DeVries, for example, suggested that mutations alone could produce new species, virtually rendering Darwinian natural selection obsolete. But most geneticists were content to focus on the details of their rapidly developing field rather than to take the time radically to rethink evolution. Those who did look at the evolutionary implications made little lasting impression on the field.

Evolutionary biology plunged into a sort of Dark Ages in the first two decades or so of the twentieth century. It was like a Tower of Babel out there, with some strict Darwinians locking horns with geneticists—and some paleontologists feeling sufficiently unfettered that they could announce their own version of heredity + evolution. The paleontologist Henry Fairfield Osborn, for example, who was president of the American Museum of Natural History, believed in a form of innate improvement within lineages that in itself leads to the emergence of superior forms. Though he thought his titanotheres (large mammals found through a succession of Tertiary deposits in the American West) evolved in this fashion (he called it "Aristogenesis"), it is no coincidence that Osborn himself was the closest thing to an American aristocrat there could be. It is also no coincidence that his name came to be linked with the eugenics movement and, worse, with the emergence of policies underlying Adolf Hitler's ethnic cleansing programs of World War II.

But Darwin's natural selection was based, not on the details of how inheritance works, but on the existence of heritable variation in local populations. The mistake was to assume, as some biologists manifestly did, that once the process of inheritance began to become clear, it alone would suffice to explain how genetic information changes through time. Not until the 1930s, when Dobzhansky spelled things out in no uncertain terms, did it become crystal clear that there are distinct levels of biological process here: a distinction between the organism level and that of entire populations. Mutation, crossing-over, and a host of other processes occur at the individual organism level. Selection—and accidental factors ("genetic drift," an idea introduced by the geneticist Sewall Wright in the 1930s)—act on variation within populations. Darwin was right, though he never actually put it this way; much as he would have loved to, he did not really have to understand the details underlying the production of heritable variation. He only had to know it was there to formulate his Malthusian idea of natural selection. The same, of course, holds true for Alfred Russel Wallace.

The Evolutionary Synthesis. By the 1920s, as laboratory techniques and experimental results became more sophisticated, mutations no longer loomed as necessarily large-scale and deleterious, thus providing the ultimate source of genetic variation in populations. And geneticists had also begun to speak of a multi-factorial (multi-genic) basis of inheritance of many characteristics, eliminating the stark dichotomy between original Mendelian either/or patterns of inheritance *vis-à-vis* Darwin's insistence upon smoothly continuous variation in populations.

It took three mathematically gifted geneticists finally to break the impasse and reconcile the new genetics with the older Darwinian vision: Ronald Fisher and J. B. S. Haldane in England, and the American Sewall Wright, writing primarily in the years just after World War I into the 1930s. They and their colleagues essentially founded "population genetics": the analysis of the fate of gene frequencies in populations given varying rates and intensities of mutation, selection, and the migration of genes ("gene flow") with other populations.

Fisher in particular restored the power of natural selection to center stage in evolutionary theory. And in a manner reminiscent of Darwin himself (though Fisher was essentially a statistician and no naturalist while Darwin rarely resorted to mathematics) he painted a picture of the evolutionary process that focused almost entirely on variation and selection. There is little in the Fisherian tradition incorporating notions of isolation, species, and speciation. Rather, evolution is seen above all as a process of adaptation—of selection tracking environmental change, and modifying the features of organisms accordingly, given the presence of sufficient variation on which to work.

This tendency to view adaptation through selection as virtually the only important process of evolution (complex and important though it undoubtedly is) dominates some quarters of evolutionary biology to the present day. The advent of molecular genetics occasioned a sort of conservative reversion to an adaptation-through-natural-selection purist position that in a sense captures the essence of Darwin at his most narrow frame of mind. Gone from such pursuits as the "selfish gene" is any serious attention to the context of natural selection: how selection acts in populations (and not over entire species—a distinction that often is missing in mathematical population genetics, just as in certain contexts it was missing in Darwin's work); the circumstances under which selection acts to effect directional adaptive change—and the circumstances when it acts to stabilize populations; and the relation of isolation and actual speciation to the generation and retention of evolutionary novelties. These are all serious issues, attracting attention from a wide range of evolutionary biologists, from molecular and ecological geneticists to paleontologists in modern evolutionary biology.* But it is entirely possible to read a book on evolution in the early twenty-first century and come away with the impression that evolution is solely a reflection of genes competing with one another to be represented in the next succeeding generation; or that all there is to the evolutionary process is the generation of evolutionary adaptive change through selection working on heritable variation within species. It simply isn't so.

Sewall Wright devoted his career to the analysis of breeding results, and to aspects of the mathematical theory of evolutionary processes. He was by no means a field naturalist; yet it was Wright who developed the picture of the genetic structure of species still very much in use today. Wright spoke of "demes," local breeding populations within species. Demes are quasi-independent parts of species, with their own subsets of the total genetic variation present over the species as a whole; with their own semi-independent mutational and selectional histories; and subject, as well, to what Wright called "genetic drift," the accidental fixation of alleles in populations that is independent of the action of natural selection.

In the early 1930s, Wright developed a graphic metaphor: the "adaptive landscape."

*Some beautiful work highlighting all these themes is summarized at the end of this chapter.

The peaks in this landscape were initially meant to be occupied by what Wright called the "more harmonious" combinations of alleles—those that produced the most robust and "fit" individuals. The problem of evolution, as Wright saw it, was to maximize the number of individuals with those most salubrious gene combinations, minimizing the less harmonious combinations.

But Wright himself—along with other geneticists, most notably Theodosius Dobzhansky—quickly co-opted the landscape metaphor for larger purposes. Demes were seen as occupying the peaks. The problem of evolution in this context became the fate of genetic information within the demes, and as that information flowed between demes. For example, how does a new, improved combination of genes, producing better-adapted individuals, and arising in one population (deme), manage to get to surrounding demes and eventually spread throughout an entire species?

Darwin and virtually all his contemporaries—including Moritz Wagner—thought that the capacity to interbreed means that species are inevitably stabilized, even downright homogenized. Indeed, that is why Darwin embraced isolation and the development of reproductive isolation in *The Descent of Man*—so sure was he that evolutionary novelties developed in one part of a species stood a good chance of being lost through crossbreeding, unless isolation severed the reproductive ties between populations with these novelties and the parental species.

As we have seen, Moritz Wagner, apparently without the benefit of any specific knowledge of the nature of the fossil record of species' histories, derived his notion of long periods of species stability strictly from his assumptions on the stabilizing power of interbreeding across species. But species are not "panmictic," with free and complete interchange of genes among all component populations over their entire range, unless a species itself consists of one or only a few localized populations (the case, for example, on archipelagos such as the Galapagos). Most species are rather wide-ranging, as Darwin himself saw when he compared the ranges of species on mainland South America with those he saw on islands. And while the common rhea shares many genes throughout its range in South America—and genetic continuity undoubtedly (if sometimes sporadically) does exist between the most far-flung populations of, for example, the common rhea over the continent—Wright's "adaptive landscape" imagery offers an alternative way of deriving the prediction that typically species will remain stable for long periods of time.

The simple fact of the matter is that species very typically do not change in the progressive, gradational manner Darwin predicted they should, unwisely adding that his theory would stand or fall on the eventual uncovering of such examples. That prediction, as we have seen, was based on his vision of natural selection modifying an entire species—no matter how far-flung—through time in response to changing environmental conditions. Wagner thought that interbreeding throughout a species would prevent selection from modifying the species further.

Paleontologists have been searching, almost entirely in vain, for these insensibly graded series. We have finally come to realize that not finding Darwin's predicted insensibly graded series points to something deeper than his conclusion that the record is hopelessly incomplete. Darwin himself acknowledged that species tend not to change even through rather thick sequences of sedimentary rock; and paleontologists—especially those working on marine fossils, whose records are characteristically more dense than those of vertebrates to begin with—now pretty much universally admit that stasis, the stability of species, is a real phenomenon. No matter how many feet of rock devoid of fossils interrupt the recorded presence of a species through an entire sequence, when fossils are next found, the species looks very much the same as it did in samples collected below—and in samples collected higher up.

We now know that invertebrate marine species characteristically remain essentially unchanged for upward of 5 or 10 million years, and sometimes even much longer. Vertebrate species—mammals or dinosaurs, for example—when known through thick sequences, remain stable for somewhat shorter periods of time, up to several millions of years, still a prodigious span of time. Clearly this is inconsistent with Darwin's rethinking of evolutionary patterns, re-derived through his thought experiments on how selection acts on species through time. Evidently stasis demands evolutionary explanation.

I bring up stasis at this point, in conjunction with Sewall Wright's analytic imagery of the genetic structure of species, because I firmly believe that Wright or someone else (a latter-day Moritz Wagner) should have been able to derive stasis as a prediction from Wright's picture of the demic structure of species. Recently, a team of geneticists and paleontologists did an exhaustive analysis of all genetic processes known to either promote or hinder evolutionary change.* None of the factors that act against the generation or accrual of genetic change—factors such as low mutation rate, thus low vari-

*See N. Eldredge, et al., "The Dynamics of Evolutionary Stasis," *Paleobiology*, 31 (2005).

Darwin's 1844 *Essay* is a more fluidly written, expanded version of the 1842 *Sketch*. The sequence of topics and much of the actual wording of this *Essay* was to appear in his *Origin of Species*, published fifteen years later. Darwin did not publish his 1844 *Essay*; but he did write a note to Emma asking her to publish it in the event of his death. He allocated £400 to cover the costs of its publication.

A colorplate from one of Darwin's barnacle monographs. The eight years that Darwin spent on barnacles resulted in work that is still consulted, and enhanced his reputation as a zoologist. But the work also filled time as he continued to avoid publishing his evolutionary views. His ideas were so well developed by 1842 that the later work on the barnacles contributed little if anything of substance to his theory of evolution through natural selection.

[*From the* JOURNAL *of the* PROCEEDINGS OF THE LINNEAN SOCIETY *for August* 1858.]

On the Tendency of Species to form Varieties; and on the Perpetuation of Varieties and Species by Natural Means of Selection. By CHARLES DARWIN, Esq., F.R.S., F.L.S., & F.G.S., and ALFRED WALLACE, Esq. Communicated by Sir CHARLES LYELL, F.R.S., F.L.S., and J. D. HOOKER, Esq., M.D., V.P.R.S., F.L.S., &c.

[Read July 1st, 1858.]

London, June 30th, 1858.

MY DEAR SIR,—The accompanying papers, which we have the honour of communicating to the Linnean Society, and which all relate to the same subject, viz. the Laws which affect the Production of Varieties, Races, and Species, contain the results of the investigations of two indefatigable naturalists, Mr. Charles Darwin and Mr. Alfred Wallace.

These gentlemen having, independently and unknown to one another, conceived the same very ingenious theory to account for the appearance and perpetuation of varieties and of specific forms on our planet, may both fairly claim the merit of being original thinkers in this important line of inquiry; but neither of them having published his views, though Mr. Darwin has for many years past been repeatedly urged by us to do so, and both authors having now unreservedly placed their papers in our hands, we think it would best promote the interests of science that a selection from them should be laid before the Linnean Society.

Taken in the order of their dates, they consist of:—

1. Extracts from a MS. work on Species*, by Mr. Darwin, which was sketched in 1839, and copied in 1844, when the copy was read by Dr. Hooker, and its contents afterwards communicated to Sir Charles Lyell. The first Part is devoted to "The Variation of Organic Beings under Domestication and in their Natural State;" and the second chapter of that Part, from which we propose to read to the Society the extracts referred to, is headed, "On the Variation of Organic Beings in a state of Nature; on the Natural Means of Selection; on the Comparison of Domestic Races and true Species."

2. An abstract of a private letter addressed to Professor Asa Gray, of Boston, U.S., in October 1857, by Mr. Darwin, in which

* This MS. work was never intended for publication, and therefore was not written with care.—C. D. 1858.

Lyell and Hooker arranged to have Wallace's paper presented to (and later published by) the Linnaean Society in 1858. A tripartite presentation, Darwin's two contributions consisted of an excerpt of the 1844 *Essay* and a portion of a letter he had written to the American botanist Asa Gray in 1857.

Darwin's experience with pigeon breeding gave him firsthand knowledge of artificial selection. Here the wild pigeon *Columba livia* (the rock dove) is surrounded by an assortment of specialized breeds of domesticated pigeons. When domesticated pigeons escape and form city flocks, they interbreed freely and show a marked tendency to revert to the "wild type." Wallace's 1858 paper argues that varieties in the wild do not necessarily revert to the ancestral condition, as is often observed among domesticated animals such as these pigeons.

A surviving page of the handwritten manuscript of *On the Origin of Species by Means of Natural Selection* (1859)—the "abstract" of his larger work already partially written, and the book that shook the world so hard its reverberations are still being felt.

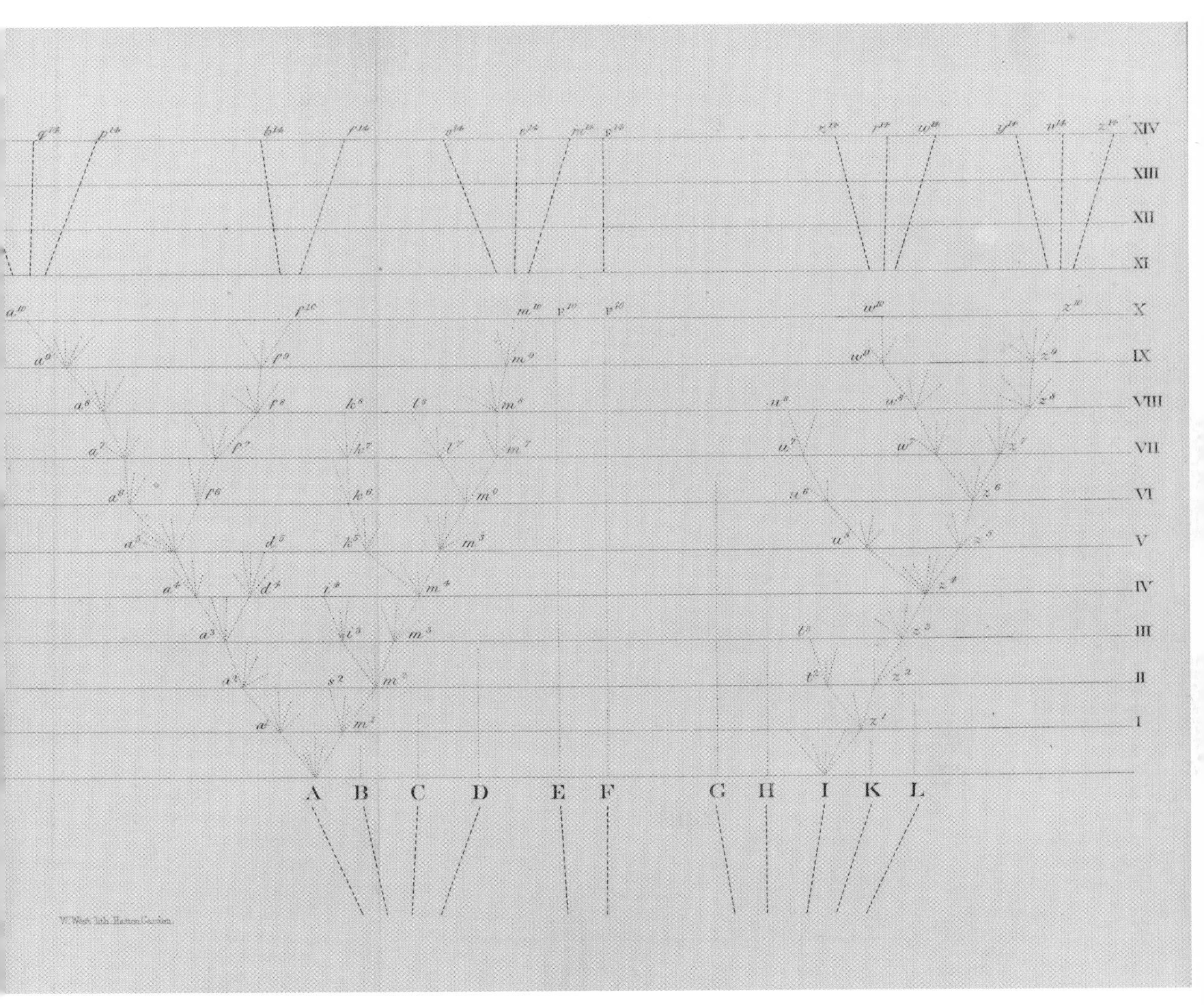

The only illustration in the *Origin of Species*. Darwin makes use of this figure several times in his text to illustrate the expected outcome of "descent with modification"—from the point of view of classification, comparative anatomy and embryology, and the lineages formed by the evolutionary process through geological time.

A caricature of Bishop Samuel ("Soapy Sam") Wilberforce in *Vanity Fair.* The famous encounter between Wilberforce and Huxley at Oxford University in 1860 crystallized the opposing positions of traditional religious creationism and the newly minted theory of evolution through natural selection.

A depiction of gradual divergent evolution of two species from a common ancestor. By the time he wrote the *Origin of Species*, Darwin had come to believe that, while isolation is "no doubt important" in evolution, large geographical expanses on continents (or in the oceans) are of even greater importance. Gradual divergent evolution was the main picture he painted, faithfully rendered in a diagram redrawn from a classic American paleontological text of the mid-twentieth century. The number of examples of this pattern of evolution documented in the fossil record are actually vanishingly small (arguably nil); testimony to the grip that Darwin's imagery had on evolutionary biology in general, and specifically on paleontology, where the patterns actually speak loudly against a slow, gradual, and progressive mode of divergent evolution.

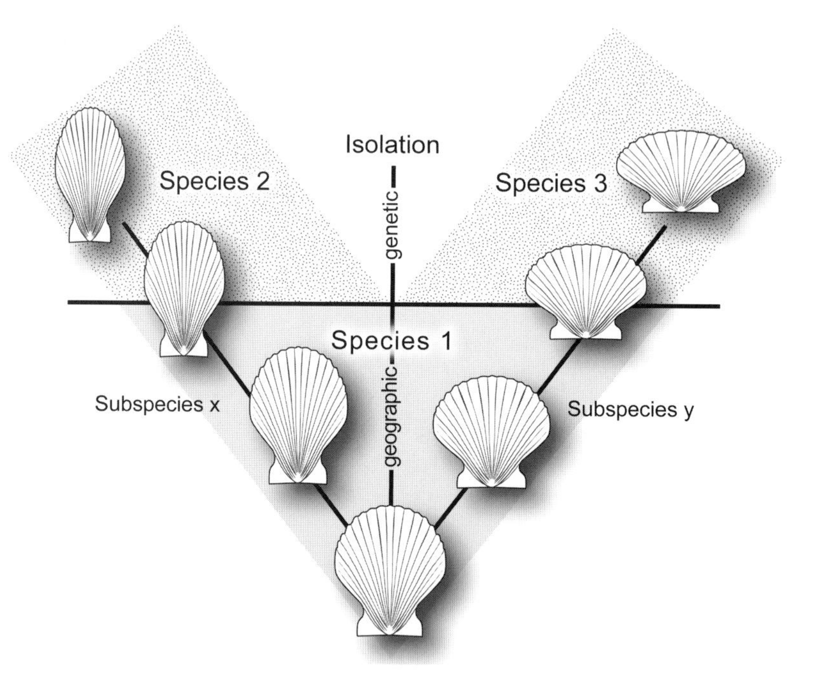

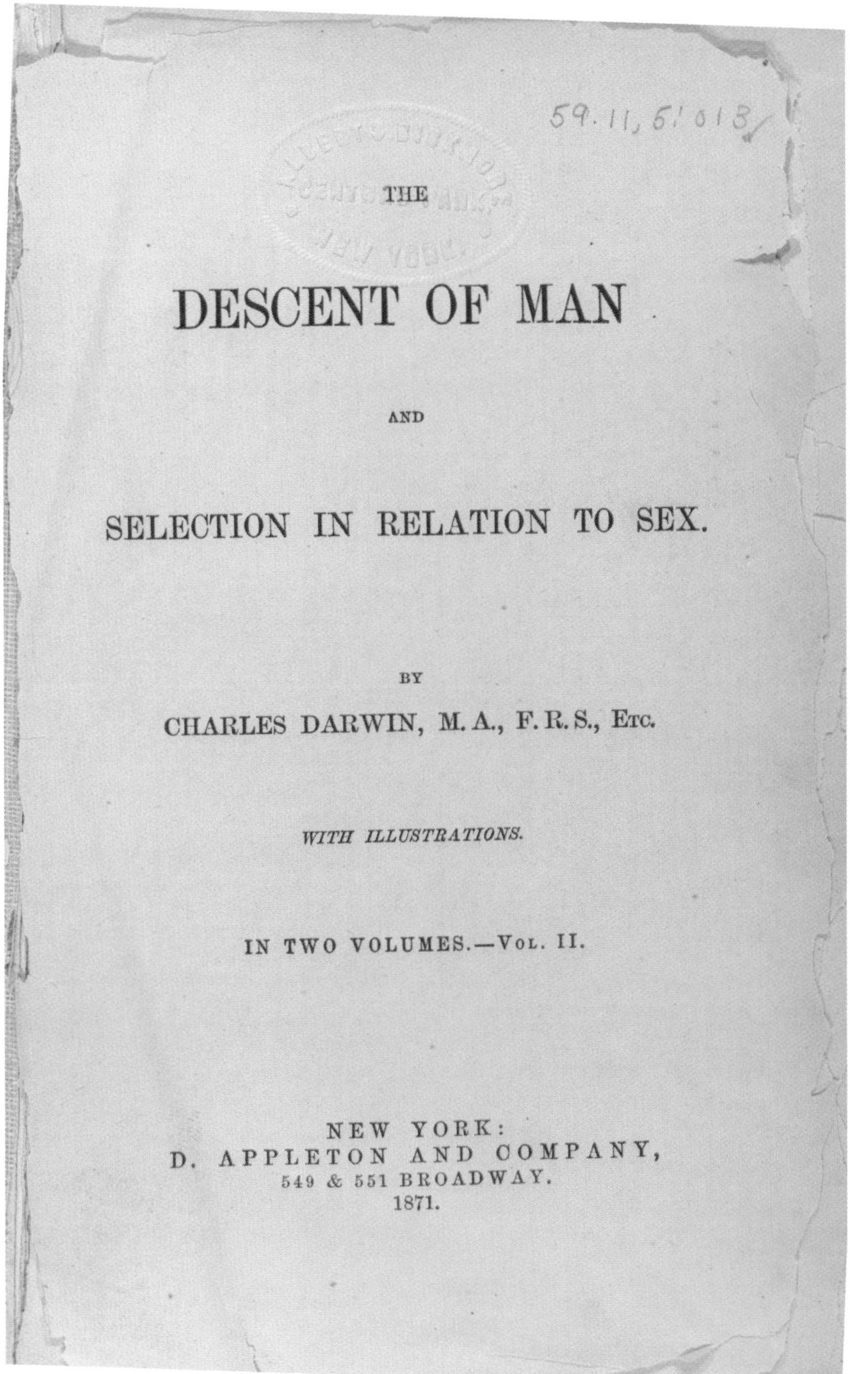

THE

DESCENT OF MAN

AND

SELECTION IN RELATION TO SEX.

BY

CHARLES DARWIN, M. A., F. R. S., Etc.

WITH ILLUSTRATIONS.

IN TWO VOLUMES.—Vol. II.

NEW YORK:
D. APPLETON AND COMPANY,
549 & 551 BROADWAY.
1871.

The title page of Darwin's last great book devoted specifically to the evolutionary process: *The Descent of Man and Selection in Relation to Sex* (1871). Although he had already concluded by the late 1830s that humans had evolved along with all of the rest of life, he merely hinted at human evolution at the end of the *Origin* in 1859, leaving this one last important task for a later time.

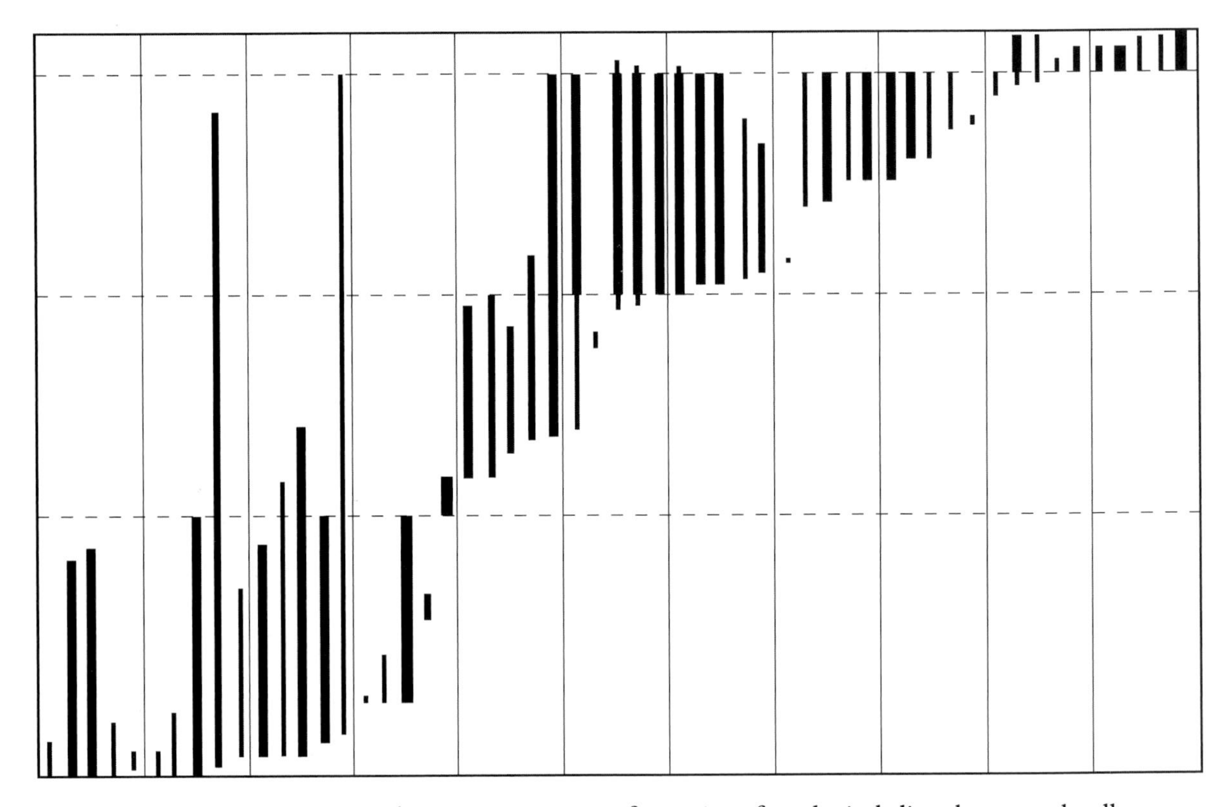

Evolutionary "turnovers." There is a spectrum of severity of ecological disturbance and collapse, ranging from localized events with little discernible evolutionary change, up through the five major mass extinctions that greatly altered the complexion of life on earth through extinction and evolutionary recovery. The mid-range, regional "threshold" level—where entire species become extinct and new species evolve to take their place—has happened thousands of times since the evolution of complex forms of life over half a billion years ago. Most evolutionary change in life's history seems to be concentrated in such events. The diagram here is redrawn from the work of the paleontologist Susan Longacre; it depicts a sequence of such turnovers of trilobites and other species in the Upper Cambrian of North America. Each vertical line represents the history of a single species.

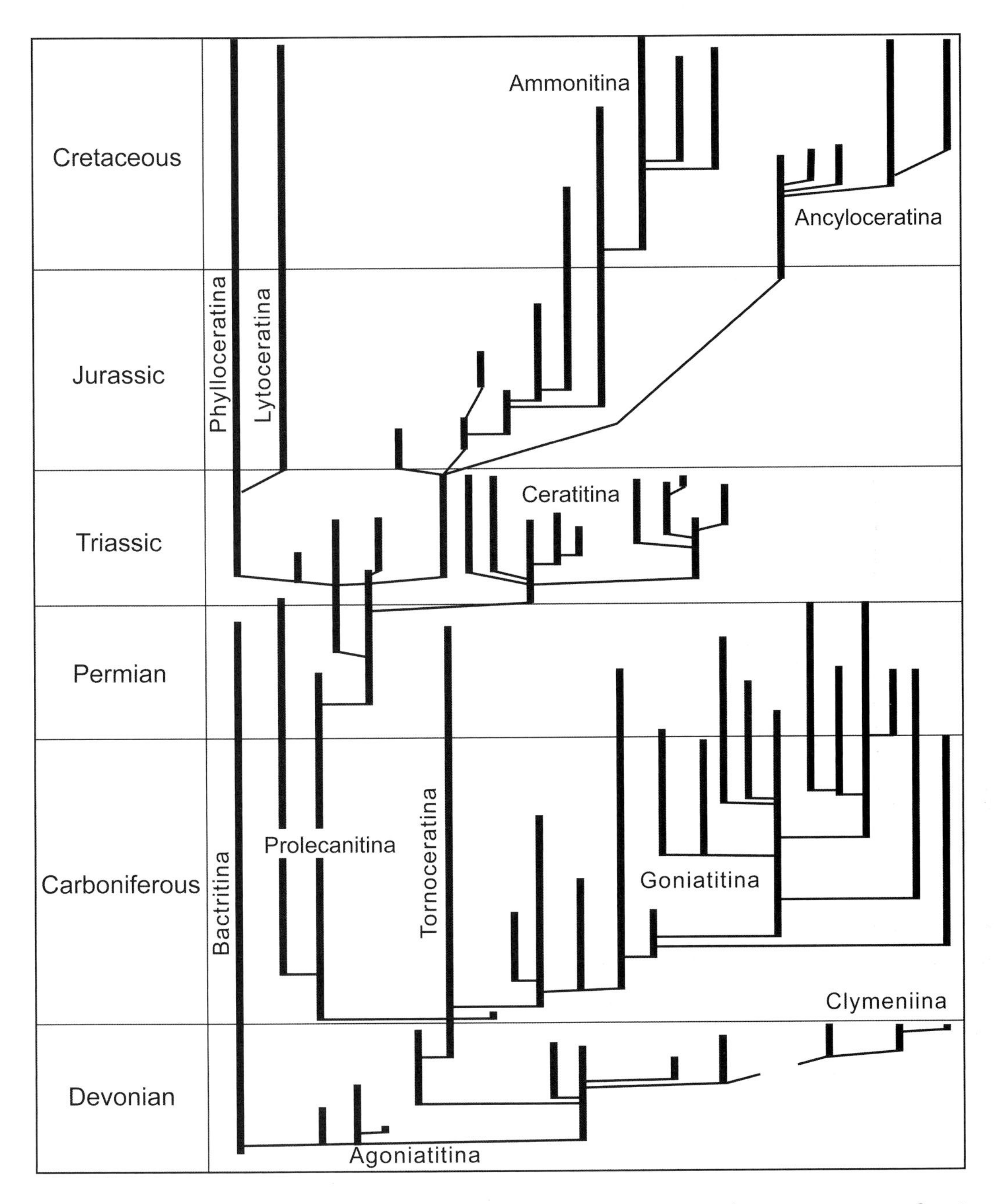

Sketch of the main features of the evolutionary history of the ammonoids, a major group of extinct cephalopod mollusks. Arising in the Devonian, the ammonoids survived three mass extinctions (end Devonian; end Permian; end Triassic), finally succumbing to extinction in the massive event at the end of the Cretaceous that is famous for driving the terrestrial dinosaurs (and many other groups) to extinction. At each major mass extinction, the dominant, most diverse group of ammonoids died out. Just a few minor lineages survived, one of which subsequently underwent great evolutionary diversification—until the next mass extinction event came along and the process was repeated.

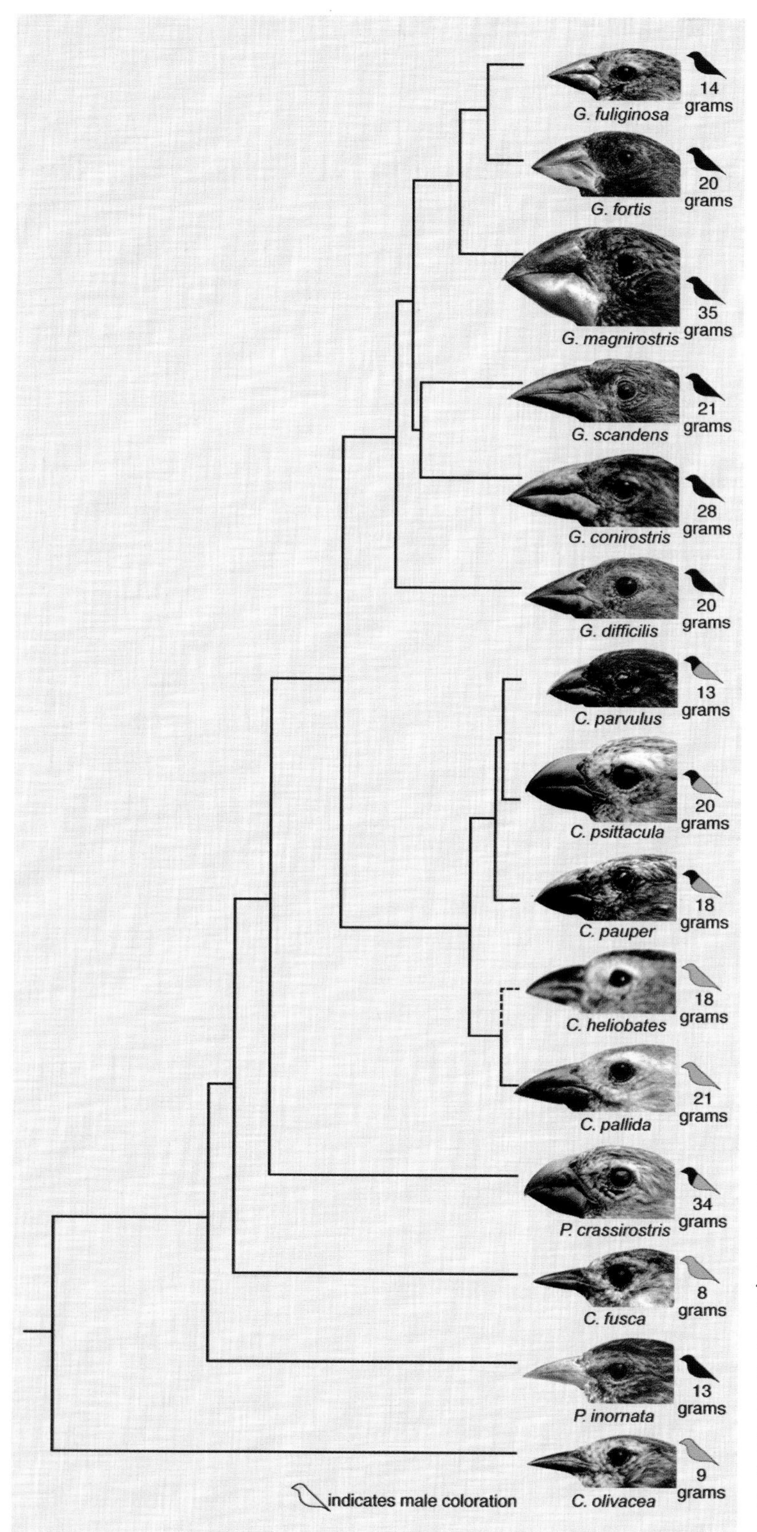

The evolutionary history and adaptive radiation of Darwin's finches on the Galapagos Islands. The diagram brings up to date the initial illustration of John Gould, published by Darwin in the second edition of the *Voyage of the* Beagle. This modern analysis, by Peter and Rosemary Grant, is based on years of field study, laboratory observations and measurements, and recent techniques of genetic analysis.

ation, or weak selection—has ever been found to have more than a transitory stranglehold on evolution. All such situations are eventually overridden, as revealed by a host of data derived from experiments and mathematical theoretical analyses.

But analyses of fossil populations, and the genetics of recent populations, reveal that localized populations indeed undergo quasi-independent evolutionary histories. Especially in situations, such as Darwin's rheas, where species populations are distributed over truly large regions of a continent (or seafloor), localized demes are almost invariably subjected to different ecological conditions: climatic conditions, including the daily and annual temperature fluctuations; rainfall and amount of surface water, which may also vary annually; nature and amount of available food; kinds of predation and disease faced by local populations; and so forth. And even when there is major climatic change (as when northern hemisphere glaciers make their way slowly southward, and then eventually retreat), the resultant pattern is not for a species to stay in place and adapt but rather to move (southward, in the case of encroaching glaciers), migrating to habitats similar to the ones to which they were originally adapted, and maintaining the same sort of patchwork quilt pattern of localized demes adapting to local conditions.

All that was needed, back in the 1930s, was for someone with an ecological bent to look at Wright's work, imagine that localized demes were inevitably faced with environments that were by no means uniform over an entire species' range, and derive the prediction that the chances of natural selection slowly changing the adaptations of species distributed so broadly over such a range of conditions must be vanishingly small. Instead, it is a "zero-sum" game: where environmental change has differing effects (including potentially migration) on the genetics of local populations, depending upon the variation present, the mutations that occur, and the selection regime that results, add up to little or no net concerted change affecting the entire species in precisely the same way.

That's what we see in the natural world: geographic variation among demes at any one time (known to Darwin, of course). Through time, patterns of geographic variation are also present, and often the entire species seems to oscillate gently, but seldom to the degree that the average state of any particular varying character shifts appreciably away from the average condition of the entire lineage—over millions of years. We'll see this pattern of oscillating stasis in more detail when we look at the Galapagos finch evolutionary story at this chapter's end.

But no one managed to emulate Moritz Wagner and predict stasis on first principles based on Wright's work. Not until the 1960s would the empirical reality of stasis—and the need for its explanation—begin to impress itself on paleontologists. But meanwhile, there was plenty that was added to the ongoing Darwinian revival of the 1930s and 1940s. The work of Fisher, Haldane, and Wright was the first phase of the new (so-called modern) synthesis: the resurrection of natural selection, now reconciled with what had been learned to that point about the nature of heredity. The second phase of the synthesis was the restoration of natural historical pattern to the evolutionary discourse. And although perhaps most of the statements were of the by-now familiar sort dictated by Darwin himself—that the patterns of botany, embryology, and so on, are all consistent with the reemerged Darwinian vision of evolution through natural selection—nonetheless some important statements of the nature of the evolutionary process did emerge in the second phase of the synthesis.

Foremost, in my view, was the resurrection of the importance of isolation, and the role it plays in generating and conserving adaptive genetic change in the emergence of new species—now once again seen firmly as reproductive communities, as independent "packages" of genetic information. For this we have to thank the geneticist Theodosius Dobzhansky, followed shortly thereafter by the systematist Ernst Mayr.

Dobzhansky was initially trained as a systematist in his native Russia. He came to New York in the 1920s to work in the already famous genetics labs at Columbia University founded by Thomas Hunt Morgan. Dobzhansky always believed that, whatever experiments (for example with fruit flies) might seem to suggest with regard to mutation and selection, we could not be utterly confident unless we can show the same sorts of processes acting in similar ways in natural populations. He spent a lot of time in the field trying to do precisely that.

Like Darwin, Dobzhansky was struck by the clear-cut differences between species living cheek by jowl in the same space. But he was also impressed by the discontinuities between species that replace each other geographically, or with only a slight overlap (like the two species of South American rheas). Dobzhansky concluded that Darwinian theory, with its emphasis on continuity, was good as far as it went, but that it did not address discontinuities between species—which Darwin had written off as another sort of artifact, not of sampling (as with the fossil record, in his view), but rather because intermediate species fell victim to extinction, and so were no longer present to fill in the gaps.

Dobzhansky thought, in contrast, that there must be an evolutionary reason for the sharp discontinuities between closely related species. He reasoned that there is, within a range, an optimum amount of variation ideally present within a species. He saw a sort of trade-off between, on the one hand, the action of selection to focus species narrowly on Wright's adaptive peaks—a process that culls extraneous variation and maximizes the adaptedness of all individuals within a species. And, on the other hand, the contrasting need to maintain variation for further evolution to occur, as when, for example, environmental conditions change. Thus isolation is necessary to maintain focus of adaptations—a point similar to Darwin's notion of the importance of isolation in conserving adaptations. Intermediates, thought Dobzhansky, would be caught in a sort of adaptive no-man's-land, vitiating the integrity of narrowly focused adaptations of two closely related but somewhat differently adapted species.

Dobzhansky concentrated on identifying the pre- and post-mating mechanisms that contribute to isolation, and thus the formation of reproductively isolated species. Mayr modified these, and contributed the famous "short definition" of biological species: "Species are groups of actual or potentially interbreeding individuals reproductively isolated from other such groups." Mayr pointed out that Darwin had never really solved (or perhaps even really addressed) the origin of species in his book of the very same title because he had focused on the wholesale generation of adaptive change through natural selection, letting new species simply emerge as adaptively modified populations. This is not (as we have seen) wholly true. And Darwin did—again, at least sometimes—clearly see the importance of isolation in generating new species, and in generating as well as conserving novel adaptations.

Thus species and speciation as topics related to (but not synonymous with) the generation of adaptive change through natural selection were firmly reestablished in evolutionary theory. They have remained alive and well ever since—despite the reversion in some quarters to a narrower focus, once again, on adaptation through natural selection, with scant regard for the environmental and geographic context vital to understanding how those processes actually work in the natural world.

Evolutionary biologists have come to abandon Dobzhansky's belief that the *purpose* served for isolation is to keep individual species focused on their adaptive peaks. Speciation is an accident, almost invariably the result of isolation. Populations cut off from the main range of their species may undergo adaptive change through natural selection—if conditions are sufficiently different from those of the rest of the ancestral range of a species, and if the requisite heritable variation is there. As Wagner, and even Darwin, observed, such change might be quite rapid, as we'll also see below in the Grants' work on Darwin's finches. It is very much as if populations, living under the less optimal conditions typical of the margins of a species range, when isolated can rapidly adapt, redefining the marginal habitat as optimal as their adaptations are honed the better to fit these conditions.

Indeed, adaptive differences (along the lines of Darwin's principle of divergence) expectedly enhance the chance of survival of a fledgling species should contact be reestablished with the parental species (as often happens on continents, but also on archipelagos). Tradition, at least since the 1940s, has it that full reproductive isolation happens largely as a byproduct of the accumulation of genetic change—most or all of which is adaptive in nature. But as the geneticist Hugh Paterson has pointed out, all that is really required for new species to emerge are changes in the purely reproductive adaptations (structures, behaviors) of organisms. Selection for continued recognition of mates might lead to the evolution of different mating signals; selection is acting just to maintain successful mating. And if, in isolation, mating systems diverge, they do so not to create isolation for some imagined adaptive purpose, but rather just as an accidental byproduct of selection for continued reproductive success. This is worth noting especially because it fits in beautifully with Darwin's distinction between sexual and natural selection (so long ignored), and also, of course, with Weismann's distinction between the soma and the germ line.

Thus there has been something of a return to Darwin's early musings on species-as-individuals. Darwin speculated as early as the Red Notebook that just as individuals have births, histories, and deaths, so too might species. After all, species do go extinct—and they do have some sort of length of stay on the planet. Might they also not have a natural process of "birth"? That's when he was toying with the idea of a "per saltum" origin of species, and thinking, as well, that there might be an intrinsic lifespan to species: that extinction, like the death of individuals, might reflect an internal, intrinsic process of senescence and death. Later, of course, he not only abandoned "per saltum" for the "birth" of species but came to see extinction as a reflection primarily of the losing battle for continued existence between older species and their more highly adapted descendants. As we'll see in a moment, most biologists today agree that the extinction of species reflects habitat destruction and loss: changes in the physical environment (whether natural or under the artificial hand of man, as is happening right now) are the prime cause of extinction.

But seeing species as individuals, stripped of their extra baggage of built-in longevities, while still recognizing that species, like organisms, do indeed have origins, histories, and deaths (extinctions), has led to a lively debate on the role that the differential generation and survival of species might have had in generating the larger patterns in the history of life. All this, too, began with Darwin, though he overwhelmingly supported the idea that long-term patterns of evolutionary change are based on generation-by-generation processes of natural selection. Darwin did see that isolation puts species in the position of being independent actors in the evolutionary drama. And he did subtitle his most famous book, *On the Origin of Species by Means of Natural Selection; Or the Preservation of Favoured Races in the Struggle for Life.*

There is something to this possibility that the differential generation and/or survival of entire species in a lineage may have something to do with the generation of evolutionary patterns—patterns that could not result from the action of natural selection alone acting within single species, as Darwin posited. Take, for example, the generation of evolutionary trends: we know, just by looking at our closest relatives among the great apes, and especially through the rich fossil record of human evolution over the past 5 million years, that brain size has increased markedly relative to body size in the evolution of the human lineage. We might imagine how selection would increase brain size (though the precise relationship between brain size and intelligence has yet to be resolved). Yet brain size shows stasis within individual fossil species of humans. If the size of the brain does not gradually get larger within the (sometimes very long) histories of individual hominid species, how do we account for the net increase in brain size—from 4.5 milliliters to 14+ milliliters? Such increase in size seems to be associated with the origin of new hominid species. And while there is no evidence that brain size was ever *reduced* in human evolutionary history (though the recent discovery of the diminutive "Floresian man," with its tiny brain, suggests at least one reversal in the overall trend to increased size), it sometimes stayed the same as new species originated. But mostly it tended to increase. The species with smaller brains became extinct—a pattern of differential survival that might actually conform to Darwin's preferred explanation of extinction in general.

Some biologists have called such patterns "species selection," while others prefer terms like "sorting"—the problem being the exact analogy with true natural selection. The point is that seeing species as individuals has expanded the range of evolutionary process theory. This can only be seen as a healthy turn of events whereby patterns in nature are once again, at long last, allowed to suggest additional sorts of processes that play roles in shaping evolutionary history. Many of these newer ideas are bound to be abandoned as more is learned. But it is good to see Simpson's suggestion for, in effect, letting nature come to us: allowing recurrent patterns of evolutionary history to come to the surface in order to help us generate and evaluate additional theoretical possibilities that will bring us closer to a complete understanding of the evolutionary process.

The Return of Cuvier. Darwin was right about the larger-scale pattern of the fossil record—that simpler forms of life appear before more complex ones, and that there is a sort of intermediate character to the fossils sandwiched in the middle of truly thick sequences of rock spanning truly prodigious chunks of geological time. The fossils of the Devonian Period appear in a very general sort of way to stand as intermediates between those of the Lower Paleozoic (e.g., Ordovician and Silurian Periods) and the Carboniferous and Permian beds that come after. In the century and a half since publication of the *Origin*, these conclusions have proven even more obvious and well supported by all the paleontological work done in the intervening years. In particular, absolutely nothing was known of the Precambrian fossil record, and the fact that it consists solely of bacteria starting about 3.5 billion years ago, until they were joined by single-celled eukaryotic forms of microscopic life a billion years later (more on this in the last chapter). Multi-cellular life wasn't to appear for yet another 1.5 billion years.

But Darwin waxed dramatic on two issues of the fossil record which it seemed to him that, if proven true, would overthrow his theory. We have already seen that the best known of these problems, stasis—the intransigent stability of most species from the time they show up in the fossil record to the time they disappear—is easily explained by a fusion of modern genetics and ecology. Indeed, I have said that stasis should have arisen as a prediction under these principles—just as, earlier, Moritz Wagner derived it as a prediction from what he and Darwin and many of their contemporaries saw as the great homogenizing influence of specieswide interbreeding. Stasis is an empirical reality, and an unsurprising one at that. It is neither an artifact of a poor fossil record nor a falsification of anything but Darwin's specific vision of how natural selection works to change entire species gradually through time.

The other problem bedeviling Darwin never fully saw the light of day in his publications. But it was there in his notation made on the text of his 1844 *Essay*—the note that says, "If species really, after catastrophes, created in showers over world, my theory false." For that is what the fossil record does indeed suggest, and what the older paleontologists saw, of whom Cuvier, with his successions of catastrophic successions followed by separate acts of re-creation of entire floras and faunas, was foremost. But now we see that Darwin essentially solved Herschel's "mystery of mysteries," giving us a naturalistic instead of a creationist view of the development of adaptive change and the emergence of new species. How then do we reconcile the pattern of "species really, after catastrophes, created in showers over world"—a very real pattern indeed—with evolution?

Easiest to grasp are the great global mass extinctions. Five are routinely recognized since the explosive evolutionary radiation of complex forms of animal life at the base of the Cambrian Period some 535 million years ago. The greatest one so far occurred 245 million years ago—an event that claimed at least 70 percent and maybe as many as 95 percent of the earth's species. This was an event that so changed the character of life on earth that geologists in the 1840s (well before the appearance of Darwin's *Origin*) adopted the term "Paleozoic" ("Ancient Life") for the fossils and rocks below this event, and "Mesozoic" ("Middle Life," also informally known as the "Age of Dinosaurs") for the fossils and rocks above. The division between the Mesozoic and Cenozoic ("Recent Life"), similarly, occurred through a catastrophe that eliminated the dinosaurs and a host of other terrestrial and marine groups of animals, plants, and microorganisms. After this last mass extinction, mammals, which had first appeared at about the same time as dinosaurs in the Triassic Period, but which had always remained relatively small, ecologically generalized, and not very diverse, literally inherited the earth that the dinosaurs had so completely dominated. Then, after a lag of some 5–7 million years, mammals began to diversify rapidly—herbivores, carnivores, and scavengers in an impressive array of sizes and shapes. It was very much as if mammals radiated into the same basic range of ecological roles that had for the previous 150 million years been occupied mainly by dinosaurs.

Such radical, global events, involving all elements of life on earth, are relatively easy to comprehend: mass extinctions reset the evolutionary clock—and life rebounds as evolution works on whatever packages of preexisting genetic information manage to survive the catastrophe. But ecological catastrophe comes in all degrees of severity and geographic scope. A hurricane can destroy much of an island, yet leave nearby islands less damaged or even entirely untouched. Whenever localized ecological damage wipes out most of the individual organisms living there, populations are eventually restored through recruitment from neighboring areas; species are indeed packages of genetic information, and seeds, larvae, or migrating adults will return to an area, often through a sequence of "successional" stages until (usually sooner rather than later) something that looks pretty much like the original ecosystem is restored.

Local ecological disturbance followed by successional rebuilding entails little observable evolutionary change, though the actual effects on genetic variation within a species might well be profound. Global mass extinctions—on the other side of the coin—involve wholesale loss of entire groups of species ("higher taxa," like dinosaurs). Evolution of new species is so rampant following these massive events that entire new lineages are formed: higher taxa such as the mammalian orders Carnivora (dogs, cats, bears, etc.), Artiodactyla (sheep, cows, antelopes, camels, etc.), Cetacea (whales and porpoises), and so forth appeared in the second wave of mammalian evolutionary diversification following the end-Cretaceous mass extinction.

So one might expect there to be some middle-range examples lying between the utter global devastation of mass extinctions and massive evolutionary diversification, on the one hand, and the ecosystem degradation and rebuilding on a more local and transitory level on the other. Given this spectrum, specifically one might expect to see examples of regional ecological collapse that entail more than the deaths of a number of individuals belonging to any number of different species—yet less than the effects of mass extinctions where entire large groups of species are eliminated. In other words, one might expect to find episodes where many different species living in regional biotas are lost to extinction.

And that is just what we do find. The finer-scale divisions of the geological time scale (many of them recognized and named prior to publication of the *Origin*) represent just such episodes. Darwin in fact alluded to them in his 1844 *Essay*, where (as we saw in the previous chapter) he remarked that geologists working on the Secondary Epoch (the Mesozoic) saw things this way, but that Lyell assures him the divisions of the Tertiary do not reflect such abrupt events. Indeed, paleontologists still debate the degree to which the epochs of the Tertiary fit this "turnover" mold—the answer seemingly being that turnovers (extinctions and subsequent evolution of replacement species) are developed in some but not all regions where Tertiary rocks are preserved.

But turnovers there are, hundreds if not thousands of them in the last 535-million-year history of life. Some paleontologists (including myself) have concluded that most of the extinction and origin of species are concentrated together in these environmentally induced episodes of coordinated extinction and subsequent speciation. It has always been assumed—literally ever since Darwin—that both extinction and speciation are like a ticking clock, with some species going extinct, and new ones evolving, more or less all the time. But, given the fact that environmental change—especially events that might lead to fragmentation, the isolation of parts of a species from other parts—is likely to affect more than just a single species living in a region, it should come as no surprise that extinction and speciation on a regional level very much tend to come in correlated bursts.

And there is cause and effect here, as well, between extinction and evolution. If Darwin for the most part thought that extinction was a result of evolution—ancestors being inferior to their descendants, thus likely to lose out in the "Struggle for Life"—we tend to see things just the other way around now. For the lessons of global mass extinction are clear: Extinction clears the way for rapid evolutionary diversification. This goes on in less dramatic fashion in the hundreds of regional turnover events that have, in fact, been the locus of most of evolution at least since the advent of complex animal life over half a billion years ago.

Does the fact that by far most origins of new species have occurred as a response to the extinction of many older species in regional ecosystemwide pulses jeopardize Darwin's ideas—as he indeed thought it would? Only in the narrowest sense. Darwin's insistence on the gradual evolution of new species on the face of it is incompatible with the long periods of stability (stasis) of most species in a regional ecological setting. Little discernible, adaptive evolutionary change accrues for millions of years, until the inevitable ecological events (such as climate change) occur that disrupt the system, drive many species to extinction, and trigger the rapid evolution of new species.

This process is wholly Darwinian, even though the combination of factors may not be precisely the way Darwin chose to lay his mature bets on the theoretical table. Climate change of sufficient scope alters habitats dramatically—driving some species to extinction, and relegating the survivors often to remnant patches of altered yet sufficiently familiar habitat in which they can persist. These surviving populations are isolated from others of the same ancestral species. And as often as not the adaptive change that separates descendant from ancestral species can happen rather rapidly, as natural selection quickly works to adjust isolated surviving populations to a modified habitat.

The ingredients are all there: heritable variation, selection, isolation. That they tend to be concentrated into bursts of extinction and proliferation of replacing species by regional environmental events is not really shocking—nor a huge deviation from pure Darwinian thinking. Rather, it is just a recasting (as Simpson told us to do) of the combination of evolutionary dynamic factors in a way to fit the empirical facts of the matter better.

Darwin, I am sure, would be intrigued. But there is even a bit more to be said here, as a number of independent thinkers are currently finding ways to integrate what we are learning about the genetics of the developmental process, the role of the genome in the evolutionary process with the pulses of extinction, and subsequent speciation so utterly characteristic of the history of life.*

How would processes intrinsic to the genome that might conceivably be related to evolutionary change—even the appearance of new species—possibly be part of a picture where extinction and the evolution of new species is going on nearly simultaneously in *unrelated lineages*? For example, Vrba has written extensively of a major turnover event in Africa that happened roughly 2.5 million years ago: an event that involved the extinction and subsequent evolution of a wide variety of antelope and other mammalian species, possibly even including species of our own hominid lineage. How can processes of change arising in the internal genome of each of these vastly different, unconnected species go through similar sorts of events at the same time?

Some of my colleagues think that heat-shock proteins, governed by the famous *Hox* genes, might be at work here. Climate change is physical, involving temperature as well as other physical variables, like rainfall. *Hox* genes are part of the regulatory apparatus, controlling the expression of genes in the developmental process. Sometimes it is the case (as it apparently is with our own species, *Homo sapiens*, and with great apes like the chimps) that differences between closely related species lie more in the switching on and off of genes found in both species in the process of development, than in

*See the Bibliography for details: Juergen Brosius, T. Ryan Gregory, Bruno Maresca, Jeffery Schwartz , and Elisabeth Vrba.

the small amount of genes that actually are different in the two species. Perhaps such environmentally sensitive components of the genome can be altered in many different species during times of extreme environmental stress.

This is speculation—but of the kind that Darwin admired and the way that ideas are juxtaposed to come up with new, potentially better explanations of natural phenomena. The ideas are testable and may well eventually be rejected. But it is exciting to see how molecular genomics and evolutionary developmental biology might actually bring to the table new ways of thinking about paleontological patterns that tell us on the large scale what the patterns in the evolution of life look like.

But consider this: If there is anything at all to the heat-shock protein scenario, here indeed is a case where the genome is responding to physical environmental change—and alterations occur. With so much extinction happening in turnover events, the probability of survival of new genotypes goes up—perhaps way up. The result is the emergence of new species not all that different from the ones they replace, though the cumulative results over enough of such events can be striking. For all our similarities, we are very different from chimps.

Replacement of one species by another over space; one species by another over time. Darwin used the word "inosculation," suggesting a sort of kissing, and, according to David Kohn, a "radical reorganization." Species do not gradually grade into one another as their geographic boundaries meet; nor do species in the fossil record grade up into their replacements.

Neither natural selection nor the patterns of replacement—the inosculations—are falsified. Both remain true to this day. But if the genetic differences between closely related species that replace one another in space and time really do boil down more to the regulatory elements controlling the switching on and off of genes in the evolutionary process, perhaps Ernst Mayr was right all along. Maybe there is a "genetic revolution" associated with speciation after all.

Evolutionary biology is alive and well early in the twenty-first century. It encompasses far more than I have been able to suggest in this brief sketch, which has been devoted to tracing out what has happened to some of the more familiar themes in Darwin's evolutionary writing: heritable variation, selection, isolation, speciation, and extinction. The chapter concludes with a précis of some modern work on variation, selection, isolation, speciation, and the generation of a large-scale evolutionary pattern which shows how essentially right Darwin was. My example is drawn from the spectacular work of Peter and Rosemary Grant—all done on that most appropriate of all examples, Darwin's finches of the Galapagos Islands.

Evolutionary Microcosm: The Triumph of Charles Darwin

It isn't exactly true that Darwin never breathed a word of "my theory" to anyone but his innermost circle until Wallace tipped his hand, they published jointly, and Darwin wrote the *Origin* in a white heat. He did occasionally drop a hint here and there—the most famous in a passage on the Galapagos finches added to the 1845 edition of his *Voyage of the* Beagle (quoted in chapter 3).

Princeton University biologists Peter and Rosemary Grant are the latest—and by far the most thorough and successful—of a string of biologists who have been drawn to the Galapagos, and especially to Darwin's finches, as they have long since come to be known. The Grants' work has become well known through their own writings, as well as through a number of books and articles written by others. But it is worth a brief revisit here if only to reiterate that the evolutionary principles Darwin was central in developing hold up to the most rigorous scrutiny and testing, through detailed field observations, measurements, and genetic analysis. The Grants' work stands as a symbol of the rich results modern evolutionary biologists, applying ecological principles and armed with the latest sophisticated tools of molecular biology, have been attaining in the past several decades.

The Grants have worked in the Galapagos for over thirty years now. The bulk of their research has taken place on the island of Daphne Major, which is some eight kilometers from its nearest neighbor. They have achieved what Theodosius Dobzhansky said we all must do: to document and analyze in nature the processes of evolution that we glimpse through laboratory experimentation. Theirs is one of the most thoroughgoing demonstrations of natural selection in the wild. They started with the basic requisite: demonstration that there is indeed variation in body size, in beak size and shape, within populations of the local species of finches on Daphne Major, and that that variation is indeed heritable. This involved measuring adults and their offspring. The result: body and beak size and shape are indeed variable—and that variation is highly heritable.

Next step, natural selection: the Grants spent so many consecutive years in the Galapagos that they were able to study the effects of prolonged drought—and the opposite circumstance of El Niño, with higher-than-average amounts of rainfall. Both climatic events changed the availability of seeds of particular size and hardness. With the drought, larger, harder seeds became dominant, and smaller birds with weaker, smaller bills died off disproportionately in the local population of *Geospiza fortis*, the medium ground finch. With El Niño, the opposite happened: smaller seeds became very abundant, and larger birds died at higher rates than smaller birds.

The Grants have shown that natural selection is a real process in nature that modifies the adaptations of organisms—and can do so in an oscillatory manner. Beak size and shape is not utterly stable, but in its oscillatory pattern of change seems not to be destined to lead to permanent change in any one particular direction. The mature Darwin might have been disappointed to hear that, over the relatively short span of thirty years (nonetheless a very long time for a single research team to keep at it in the field), no real gradual change seemed to be accruing.

But of course there *is* adaptive differentiation among the thirteen species of Geospizine finches (there is one additional species on Cocos Island 425 miles northeast of the Galapagos). The Grants have been able to document all phases of the process of speciation. They were able to witness the establishment of a breeding population of the large ground finch on Daphne Major—a population that has diverged somewhat from its ancestral condition through adaptation to slightly different ecological conditions than on its source island (Santiago), and through genetic drift and the effects of inbreeding as well. But the most significant change, initiated by a single male, came in the song that is used in mate recognition.

Changes in mate recognition systems are necessary—and sometimes sufficient—to lead to full reproductive isolation. The Grants recount an additional study, this time documenting the near-end phase of the speciation process. Cactus finches and medium ground finches are distinct species that can, nonetheless, occasionally hybridize. Though closely related, they differ in beak size and shape—and in their mating songs. The Grants found that hybridization occurs when females mistakenly learn the mating song not from their fathers but from another nearby male of the other species. Hybridization is only possible because the two species, despite their differences, are still sufficiently alike genetically. The final stage of speciation comes when, mistake in song recognition or not, hybridization has become physically impossible.

Thus Darwin's initial hunch on the importance of isolation from island to island in the Galapagos is abundantly confirmed: new species arise there by the haphazard, yet in the long run fairly regular, occurrence of populations of different species managing to invade different islands successfully, with somewhat different ecologies, there to undergo their own adaptive diversification. This, nonetheless, is not necessarily going to survive the onslaught of inbreeding with additional migrants from the parental population unless and until there is a change in the mate recognition systems—meaning, in the case of the finches, primarily in their songs.

But there is more. The Geospizine species are indeed all descended from a common ancestor, now known from genetic analysis to have been a member of the seed-eaters, a group related to tanagers currently living in South and Central America and on islands in the Caribbean. They arrived in the Galapagos (as, again, DNA analysis suggests) between 2 and 3 million years ago. The Grants confidently conclude that the ongoing process of adaptation through selection as a response to environmental change—and the process of speciation through isolation as a way of making permanent the otherwise oscillatory changes that seem to be happening constantly—has continued for those 2–3 million years, and produced the adaptive radiation we see in the finches today.

There is little to suggest what role, if any, extinction and perhaps even focused turnovers may have played in the evolution of Darwin's finches. The last great event that occasioned turnovers—a global cooling event some 2.5 million years ago—may have been the cause of the ancestral finch showing up in the Galapagos in the first place. And the Grants feel that the pace of evolution has been accelerated in the later stages of this evolutionary history, probably reflecting the facts that there are more islands now than there were when the finches first got there, and that the sea level has oscillated radically as glaciers have alternatively grown (locking up seawater) and shrunk. No doubt about it, the environment has been the ultimate driving force behind the evolution of Darwin's finches.

But the larger point is this. With all the rich array of new techniques that lets us probe the depths of genetic structure—tools that allow us to get to the heart of variation and selection, and to reconstruct with pinpoint accuracy the sequence of evolutionary events and even their place in geological time—the same basic ingredients of the evolutionary process outlined by Charles Robert Darwin turn out to lie at the heart of today's most sophisticated versions of evolutionary theory. At one time or another, he basically saw it all. And if the version he adopted of "my theory"—the version that was colored so deeply by his vision of how natural selection works in the wild—

strayed too far from seeing the importance of isolation, or the role of extinction, or the stability of species in geological time, nonetheless he had most of it right. We may know a lot more now, but our theory of the evolutionary process is still basically his theory. And that's why, scientifically, Darwin holds a place in the modern world occupied by few other notable figures of the nineteenth century.

But there are other reasons why Darwin's name continues to be called in modern life—and some of these are not as pleasant to behold.

CHAPTER 6

Darwin as Anti-Christ: Creationism in the Twenty-first Century

When Darwin was born in 1809, the received Church of England interpretation of Genesis pretty much dominated British thinking on the origin and age of the earth, of life—and above all else, of mankind. This was as true of the "men of science" as it was of the general populace. Not only was there no real salaried class of academic scientific professionals, but many of those who were engaged in scientific pursuits (and there were many among them who were great scientists) were also clergymen.

True, rationalism was making inroads in science as well as in related pursuits such as philosophy. But the course taken on the whole was from the outside in: it was far easier to apply cold logic to problems such as gravity (with falling apples posing the metaphorical pattern begging for an explanation) than it was to conceive of organic bodies themselves exhibiting a kind of historical motion. And though some before him had proclaimed the interconnectedness of life, it was Darwin who saw that the replacement of extinct species by new ones (Herschel's "mystery of mysteries") meant that life must have evolved. Added to the fossils were the replacement of one species by a "closely allied" species geographically (his South American rheas) and the smaller-scale replacement of closely similar species on different islands in the Falklands and especially the Galapagos (at first the mockingbirds and tortoises, later the plants and the celebrated finches). Later, in London, he realized the significance of the nested pattern of homology underlying the Linnaean hierarchy, as well as embryological patterns of closer resemblance the earlier stages of embryos being compared. All these patterns were the equivalent of the Newtonian apple: recurrent patterns seen everywhere, throughout geological history, in every region of the globe, and in all forms of life.

Coupled with his discovery of natural selection in 1838, and with his masterful stroke of converting his patterns into predicted outcomes of the evolutionary process (thus rendering the idea of evolution testable in a modern scientific sense), Darwin was able to show in the *Origin of Species* that his patterns—like Newton's metaphorical apple—could have only one meaning: that all life has descended from a single common ancestor living in the remote geological past. In other words, Darwin demonstrated without a shadow of doubt that life has evolved.

And, as we have seen, he immediately converted the emerging class of salaried professional biologists to evolution. In the good tradition of all human argumentation, those who resisted his views were largely his elders. If biology hasn't bought absolutely everything Darwin had to say about evolution, in the main, everything we have learned about the molecular basis of heredity, and all the new discoveries of fossils and living species, either corroborate what Darwin originally said or comfortably fit right in as extensions of his original theory. And if some of his conclusions appear to have been incorrect—such as my paleontological discomfort with his dismissal of patterns of stasis, rapid change, and turnovers in the fossil record of the history of life—these too are not the terminally fatal objections to Darwin's theory he feared they might be. Rather, they too fit right in if isolation, speciation, and the threshold effects of physical change and extinction are incorporated into his view of evolution. It is indeed ironic that, as some passages in his notebooks and early manuscripts clearly show, Darwin even saw the way out of these difficulties. But he chose not to pursue such lines of thinking in deference to his strategy to derive his patterns directly and afresh from his vision of the workings of natural selection.

Darwin's main error, in a nutshell, was to see natural selection working on entire species through time as directly analogous to the selective breeding done by small groups of farmers, horticulturalists, or animal breeders—a mistake perpetuated to this day in some quarters of evolutionary biology. And as against that rather understandable gaffe, there is the astonishing array of thoughts and suggestions, never developed in full by Darwin, that took a century or more to come to the fore in evolutionary thinking—the "founder effect," for example, and perhaps especially his principle of divergence—among the several examples highlighted along our tour of Darwin's Red and Transmutation Notebooks, his 1842 *Sketch*, and his 1844 *Essay*.

As we have seen, in his later years—specifically, in his *Autobiography*—Darwin all but admits that he was so thoroughly enamored with natural selection that he came to see it as "my theory" when all along, and in the end as he himself acknowledged, his most important task and crowning achievement was simply to convince the thinking world of the truth of evolution. In this he succeeded very well, transcending the world of professional biology and related disciplines, convincing most thinking persons of all walks of life of the essential truth of evolution. And this included most religious people—at least of the Judeo-Christian religions as practiced in Europe and its colonies, including the United States.

Most mainstream Protestant, Catholic, and Jewish people to this day have had little or no problem with the concept of evolution. Indeed, most people in their workaday lives give as little thought to evolution as they do to plate tectonics, quantum mechanics, or special relativity—to name three other prominent scientific theories that have come down the pike since 1859. In terms of mainstream Judeo-Christian doctrine, the usual path takes shape as a rendering-unto-Caesar trade-off. Most adherents to what might best be called "moderate" religious views (not to impugn the true passion but meaning simply the less doctrinaire and extreme of the spectrum of variation in Judeo-Christian beliefs) see the job of scientific cosmology—issues pertaining to the origin and history of the universe, our solar system, our earth, and all of life, including ourselves—as the proper subject matter of science. From this point of view, the two and a half versions of cosmology in Genesis are interesting for their insights into the views on such subjects entertained by farmers living in the Middle East six thousand or so years ago. But they are not to be taken as literal insights into the truth of such matters.

Papal doctrine and the teachings in most Protestant and Jewish theological seminaries have for the most part taken this view ever since Darwin released his intellectual thunderbolt. From this angle, religion and science occupy essentially separate domains: science describes the material content of the universe and the interactions among its parts, while religion is occupied with matters spiritual. Such is the view not only of many people of religious faith but also of such scientists as my late colleague Stephen Jay Gould, who saw science and religion as "non-overlapping magisteria."

Others of religious faith have gone a bit further, actively seeking a reconciliation of science and religion. A common approach, for example, has been to interpret the Days of Creation in Genesis metaphorically—where the story of Creation is mapped on to the geological time scale. But such attempts, however personally satisfying they may prove to be, inevitably lead to intellectual trouble, as the long pre-Darwinian history of such ventures shows that there are as many specific forms of scientizing apologetics as there are apologists, and none bear close scrutiny if the analogies between creation days and geological eras are taken the least bit beyond the simplest of hand-waving metaphors.

Fundamentalist and evangelical Protestants have always been in the forefront of the unremitting creationist offensive. But, increasingly, Judeo-Christian doctrine and practice in the United States has seemingly en masse become more conservative—especially in the early days of the twenty-first century, as the battle lines are once again being drawn between the Western and the Muslim world, where armed, financed, and organized resistance to globalization is most concentrated. Much of the shift to the political right has walked lockstep with a deepening turn to more conservative, fundamentalist religious positions and "values."

Creationists if anything become more inflamed by theistic evolutionists, who perceive a disconnect between the Bible and science, but attempt to forge connections with their metaphors, than by scientists—who either simply insist that science cannot by the very rules of its procedure deal with the supernatural, or who agree with them that there is an inherent conflict between science and religion that cannot be reconciled except by the acceptance of one or the other.

Creationism persists for, I think, two basic reasons. One is simply that many people of faith insist that the Bible must be true "in all its offices," so that to call in question *anything* written in the Bible is to impugn the truthfulness of the entire document. Faith, to this fundamentalist perspective, absolutely hinges on the inerrancy of the Bible.

The more subtle position is that the moral basis of human behavior depends on what we think about who we are—and that who we are depends very much on how we came to be. If we are created in God's image, then we can aspire to be like Him and behave in a moral manner. If, however, we are "descended from monkeys," there is no distinction between us and the rest of creation—and no reason to expect us to lead a moral life (though what is immoral about a monkey I cannot imagine; monkeys may not even be amoral). In this line of thought, the possibility that morals, ethics, and laws to govern our social behavior are necessary simply for societies of our sentient, social species to function is just never entertained.

There may be other factors contributing to the persistence of creationism. But, whatever the pretext given, creationism is not a valid intellectual enterprise the way it was in Darwin's day, when the best minds in England (including the younger Darwin) and elsewhere actually embraced creationist views, finding them satisfying and intellectually compelling. The fact that the creationist battles in the United States and elsewhere are fought in the political arena—particularly in the form of attempts to change biology teaching, along with other elements of the curriculum deemed at odds with fundamentalist Christian "values"—means on the face of it that creationism has an underlying political agenda. Creationism fights a pitched battle for the hearts and minds of our young people.

The United States, amazingly yet thankfully, remains at or near the forefront of the bulk of the world's scientific research. But our efforts to remain in a leading position in science are being relentlessly hammered by creationists, who insist that they just want to give equal time to an equally valid alternative theory ("Intelligent Design" being the latest version), when in point of fact what they really want to do is get their own version of religious truth into the public school arena. Our capacity to produce more scientists is hobbled by their efforts. Worse, we are in danger of having an even more scientifically illiterate electorate than ever before—in an era when so many public, political issues (such as stem cell research) call for the considered judgment of an informed body politic.

I for one am with those scientists who insist that science is about figuring out the nature of the material universe. Its procedures involve the inductive, intuitive grasp of the nature of a problem; that's how the questions are raised, and hypotheses and theories are formulated in the first place. In that, science is like any other creative human endeavor. But—and this is a large "but"—science operates by formulating predictions about what we would expect to observe if a hypothesis is true. Observing the predicted consequences does not prove the truth of a proposition; but failing to observe the predicted outcome does call into question, perhaps even "falsify," a scientific idea.

We should be forever grateful to Darwin for doing just that with his notion of evolution through natural selection: he turned his intuitive patterns that led him to evolution into natural consequences—expectations—of what we should see if evolution is true. So the notion of evolution is testable. It has been tested so many times: whenever drug resistance is studied in medicine; when mathematical and laboratory experiments tackle small-scale patterns of stasis and change in genetic information; or when paleontologists document the relationship of extinction and evolution—whether on a single-lineage scale (Herschel's "mystery of mysteries" again) or on larger scales of regional and global turnovers; whenever a systematist finds a new species and finds its place on the diagram of relationships of previously known species; whenever a geneticist sequences DNA of a species and finds the close similarity of that DNA and that of other species already predicted to be close relatives . . . and on and on. When a theory is tested, continually, day after day in routine research in so many disparate fields after 150 years, it remains true that in some arcane logical sense we still cannot say that the theory is *proved*. But failure to falsify the basic notion of evolution after so much effort means that we are unlikely to disprove it in the future.

So evolution is abundantly "corroborated." It is every bit as much a solid theory as plate tectonics (known in its earlier guise as "continental drift," the theory of a mobile earth), quantum mechanics, special relativity, and all the other cornerstones of modern science. And we have Charles Robert Darwin to thank for not only formulating the idea but also giving us the tools for understanding the fundamental testability, hence fundamental scientific nature, of evolution.

When confronting creationists—those who, whether they admit it or not, are religiously and politically rather than scientifically or otherwise intellectually inspired—I especially stress three predicted points that arise from the simple notion of evolution. One is the nested pattern of resemblance that links up absolutely all living things (Darwin's classification point, arrived at early on after opening up his Transmutation Notebooks). The second is Darwin's point about the large-scale record of the history of life. The third was basically unknown to Darwin except as a prediction: the fossil record of *human* evolution, buttressed and amplified by the recent findings of genetics. In brief, these are the three major predictions and findings:

1. *All life, fossil and recent, is connected by a single nested set of resemblances.* As Darwin said, this must be so "on my theory"—and had already been discovered to be so by Linnaeus and other early biologists who in fact were creationists. If all life has indeed

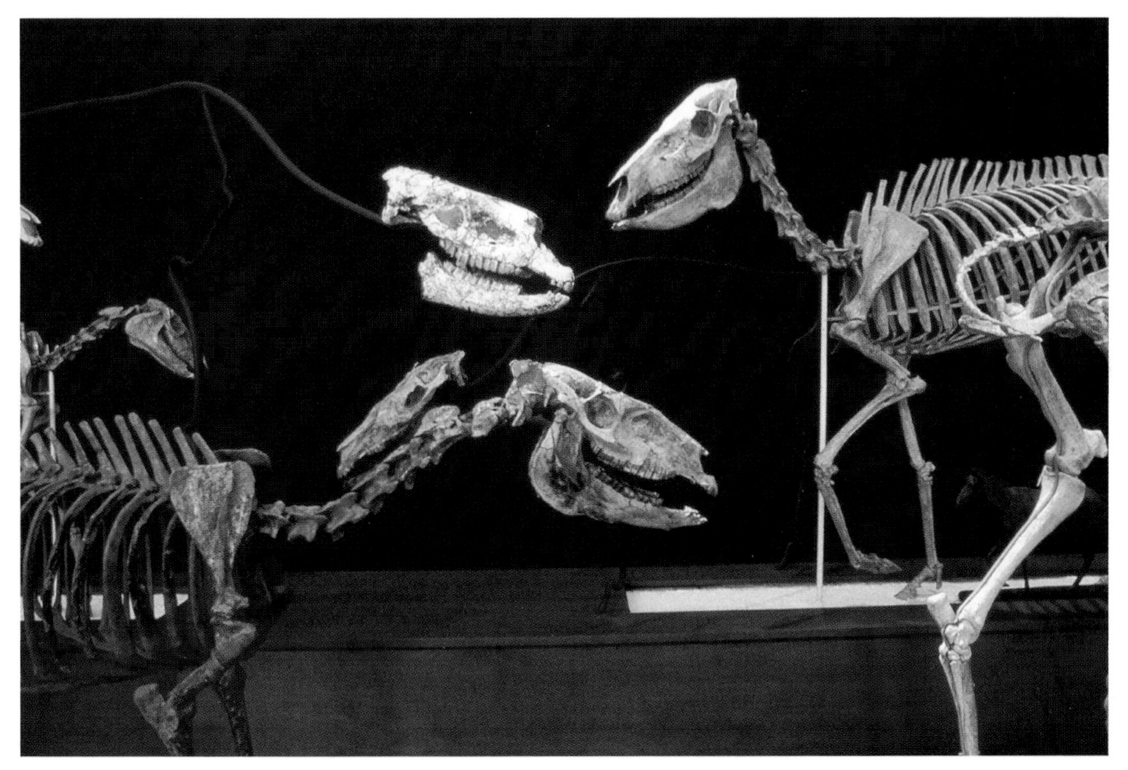

A vignette of some fossil horse species. In addition to reduction in toe number, horses became larger, their faces more elongated, and their teeth larger, more complex, and capped with "cement" when some evolved from the ancestral diet of leaf browsing to grass grazing. The cement and large size of the teeth were adaptations to cope with the silica in many grasses that grind teeth down. Eventually the browsers became extinct.

descended from a single common ancestor, we would expect there to be a trace—something shared by all of life living today passed down and inherited by all living things. Darwin could not have known what that something might be but now we do: the macromolecules of heredity DNA and RNA.

Bacteria were not known in the early nineteenth century—and it was long after their discovery that microscopists discovered that bacteria have no nucleus surrounding their single-stranded DNA; larger microbes, plus all plants, fungi, and animals, have a double-walled nucleus surrounding their DNA-bearing chromosomes. Thus these eukaryotes are descended from a lineage of bacteria (prokaryotes; the biologist Lynn Margulis has actually demonstrated that eukaryotes arose from fusion of two different prokaryotic lineages), and their cell structure bears the stamp of their common descent.

And so on. Animals share certain genetic sequences with fungi—making them closer relatives of each other than either is to plants. Within the animals, the vertebrates share developmental features with echinoderms (starfish, sea urchins, etc.)—making us closer relatives to starfish than we are to insects or crustaceans. The current understanding of the nested sets of shared resemblance—genetic, anatomical, and developmental features—and the diagram of evolutionary relationships that these resemblances demonstrate, is shown on page 226.

2. *The history of life should proceed from the simple to the complex—and this sequence should be preserved in the fossil record.* Evolutionary biologists from Darwin on down have said that the origin of life is a separate problem from what happened to life—its evolution—once it was up and running. Life's origin is a biochemical and geochemical problem whereas evolution is a matter of variation, selection, speciation, and all the other factors we have encountered so far.

We assume life originated here on earth; even if it were eventually to be shown that life arose elsewhere and spread to earth, we still assume it to have originated from non-biological (abiotic) elements. And, either way, the chemical origin of life—assembly of the first amino acids into peptide chains and proteins; assembly of the first RNA and DNA molecules—would have to entail the most simple of biological structures, simpler still than any known bacterium. In other words, we do not predict that the oldest forms of life were elephants or redwood trees. They should be bacteria or something even smaller and simpler. Creationists, in contrast, believe life was all created about the same time—except for humans, who came a day later.

In Darwin's day, the oldest known fossils came from Sedgwick's Cambrian System, then known only from Wales ("Cambria") and a few places in Scandinavia and now known the world over. The rocks below the Cambrian seemed utterly devoid of fossilized forms of life. Indeed, it has only been in the last half century or so that the fossil record of the so-called Precambrian has come to be known at all. It is remarkably extensive—and together with what has been learned of the development of complex forms of life over the past half-billion years, a remarkably detailed picture of the history of life has emerged, one that agrees very well with the predicted sequence of primitive-to-derived forms of life.

The earth is very old—some 4.65 billion years old. Some (by no means all) creationists even acknowledge this extreme old age, for to deny it is to deny radioactive dating, and hence nuclear physics and its theory of the atom and atomic decay. The oldest fossils are, as predicted, bacteria (simple tiny rods and spheres); they are 3.5 billion years old, the age of the oldest sediments yet found that could possibly preserve fossils.

The next step, one would predict from the tree of life, would be the evolution of complex cell structure—single-celled eukaryotes. Sure enough, the oldest eukaryotes are around 2.2 billion years old, meaning that, so far as we know, bacteria were the only forms of life on earth for over 1 billion years.

The simplest forms of animal life are the next to appear—but not until some 650 million years ago. More complex forms of animal life follow later, in what creationists love to pounce on as evidence of simultaneous creation of all life: the Cambrian Explosion. And it is indeed true that the more complex forms of animal life—arthropods, mollusks, annelid worms, brachiopods, echinoderms, and chordates—all do seem to appear at very nearly the same time. This is not "all life" by any means, but still it is a near-simultaneous appearance of many of the more complex animal phyla.

In fact, the Cambrian Explosion is mirrored in the lack of resolution of the genetic and other data used to construct the Tree of Life: there is no telling if the arthropods, for example, are more closely related to the mollusks than either is to the echinoderm/chordate lineage. So, we have two independent lines of evidence pointing to rapid evolutionary diversification of the "higher" animals, thought to be triggered by a rise in oxygen content in seawater to a point sufficient to support multi-cellular, multi-organed complex animal life.

The Cambrian Explosion, in other words, is no real comfort to the "all life created at the same time" position, given the sequential appearance of bacteria, eukaryotic microbes, and simple animals (like sponges and coral relatives), all separated by a bil-

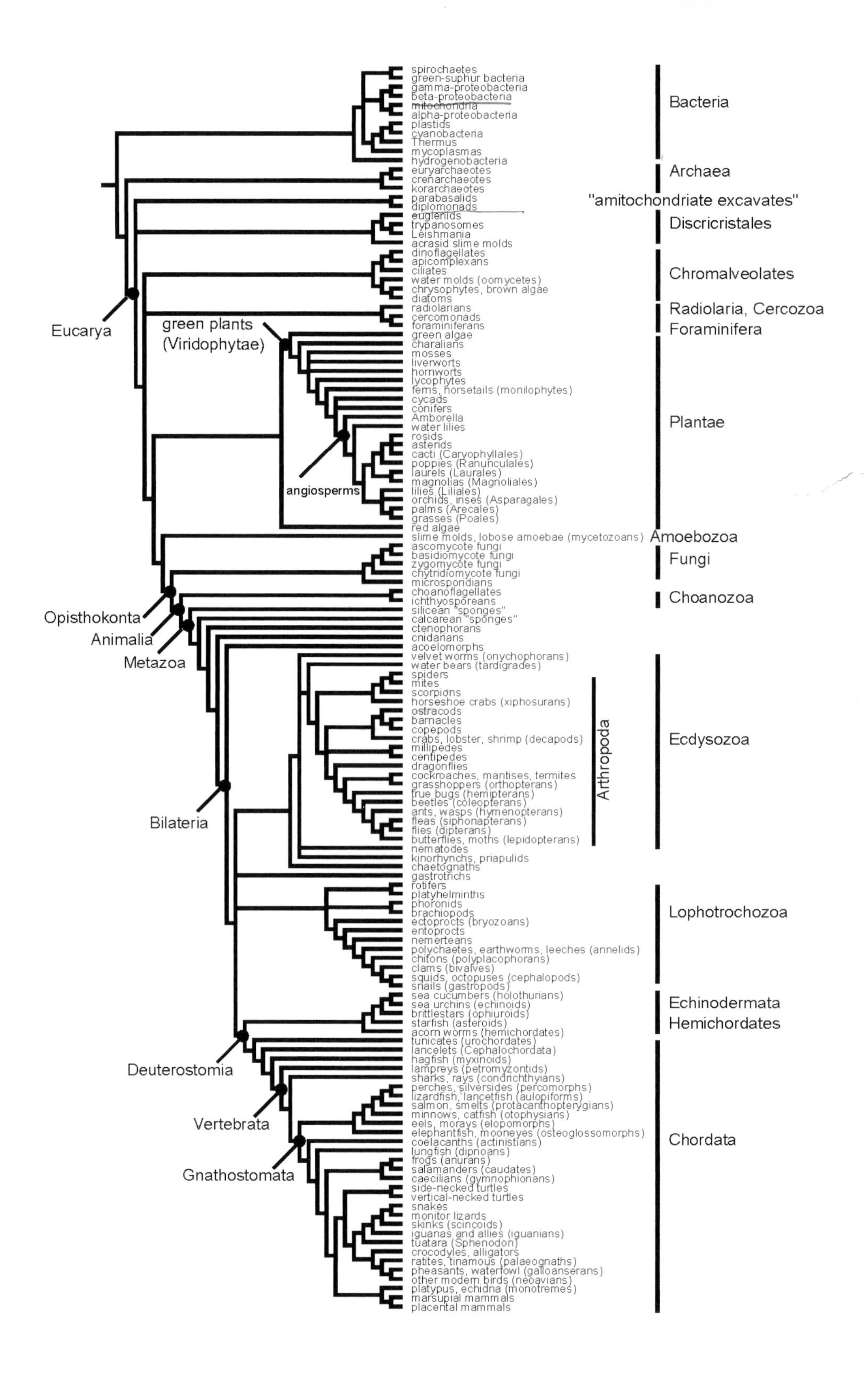

spirochaetes
green-suphur bacteria
gamma-proteobacteria
beta-proteobacteria
mitochondria
alpha-proteobacteria
plastids
cyanobacteria
Thermus
mycoplasmas
hydrogenobacteria
euryarchaeotes
crenarchaeotes
korarchaeotes
parabasalids
diplomonads
euglenids
trypanosomes
Leishmania
acrasid slime molds
dinoflagellates
apicomplexans
ciliates
water molds (oomycetes)
chrysophytes, brown algae
diatoms
radiolarians
cercomonads
foraminiferans
green algae
charalians
mosses
liverworts
hornworts
lycophytes
ferns, horsetails (monilophytes)
cycads
conifers
Amborella
water lilies
rosids
asterids
cacti (Caryophyllales)
poppies (Ranunculales)
laurels (Laurales)
magnolias (Magnoliales)
lilies (Liliales)
orchids, irises (Asparagales)
palms (Arecales)
grasses (Poales)
red algae
slime molds, lobose amoebae (mycetozoans)
ascomycote fungi
basidiomycote fungi
zygomycote fungi
chytridiomycote fungi
microsporidians
choanoflagellates
ichthyosporeans
silicean "sponges"
calcarean "sponges"
ctenophorans
cnidarians
acoelomorphs
velvet worms (onychophorans)
water bears (tardigrades)
spiders
mites
scorpions
horseshoe crabs (xiphosurans)
ostracods
barnacles
copepods
crabs, lobster, shrimp (decapods)
millipedes
centipedes
dragonflies
cockroaches, mantises, termites
grasshoppers (orthopterans)
true bugs (hemipterans)
beetles (coleopterans)
ants, wasps (hymenopterans)
fleas (siphonapterans)
flies (dipterans)
butterflies, moths (lepidopterans)
nematodes
kinorhynchs, priapulids
chaetognaths
gastrotrichs
rotifers
platyhelminths
phoronids
brachiopods
ectoprocts (bryozoans)
entoprocts
nemerteans
polychaetes, earthworms, leeches (annelids)
chitons (polyplacophorans)
clams (bivalves)
squids, octopuses (cephalopods)
snails (gastropods)
sea cucumbers (holothurians)
sea urchins (echinoids)
brittlestars (ophiuroids)
starfish (asteroids)
acorn worms (hemichordates)
tunicates (urochordates)
lancelets (Cephalochordata)
hagfish (myxinoids)
lampreys (petromyzontids)
sharks, rays (condrichthyians)
perches, silversides (percomorphs)
lizardfish, lancetfish (aulopiforms)
salmon, smelts (protacanthopterygians)
minnows, catfish (otophysians)
eels, morays (elopomorphs)
elephantfish, mooneyes (osteoglossomorphs)
coelacanths (actinistians)
lungfish (dipnoans)
frogs (anurans)
salamanders (caudates)
caecilians (gymnophionans)
side-necked turtles
vertical-necked turtles
snakes
monitor lizards
skinks (scincoids)
iguanas and allies (iguanians)
tuatara (Sphenodon)
crocodyles, alligators
ratites, tinamous (palaeognaths)
pheasants, waterfowl (galloanserans)
other modern birds (neoavians)
platypus, echidna (monotremes)
marsupial mammals
placental mammals
Bacteria
Archaea
"amitochondriate excavates"
Discricristales
Chromalveolates
Radiolaria, Cercozoa
Foraminifera
Plantae
Amoebozoa
Fungi
Choanozoa
Ecdysozoa
Arthropoda
Lophotrochozoa
Echinodermata
Hemichordates
Chordata
Eucarya
green plants
(Viridophytae)
angiosperms
Opisthokonta
Animalia
Metazoa
Bilateria
Deuterostomia
Vertebrata
Gnathostomata

lion years or more. After the explosion, a more measured pace of evolution yields the predicted pattern for all to see: in vertebrates, for example, fish appear before amphibians; amphibians before reptiles; and mammals and birds much later still—from different lineages within the reptiles. Primates are rather a primitive, early order of mammals, appearing in the Cretaceous while terrestrial dinosaurs were still around. The apes showed up some 30 million years ago—and our own human lineage diverged from the rest of the apes only about 5 million years ago. Our own species, *Homo sapiens*, is a relative newcomer, evolving in Africa roughly 150,000 years ago according to both fossil and molecular genetic lines of evidence. Which brings us to our third prediction.

3. *The fossil record should reveal a progression of hominids from more apelike, smaller-brained individuals to larger-brained, bipedal and tool-making species—up to and including ourselves. Genetic data should reveal greater kinship between us and the apes (such as chimpanzees) than between humans and any other form of life.* Both predictions are, of course, abundantly verified. Deny it though they may, all creationists *really* care about is human evolution.

Darwin immediately concluded that humans must have evolved right along with the rest of life—and said as much shortly after opening his Transmutation Notebooks. He later concluded that our resemblances lay closest with the African apes, and so predicted that Africa would prove to be the cradle of human evolution—a prediction that has been amply borne out by the profusion of fossil hominids collected there over the last half century. But during his lifetime, next to nothing was actually known about human fossils, let alone, of course, anything about genetic resemblances between humans and apes.

We know a lot more now. Early in the twentieth century there were still few enough remains of fossil humans that it was possible to line them up in a sequence through time in good linear "Darwinian" fashion—detailing, true to prediction, the progression of increasingly larger-brained species. From midcentury on, however,

A recent analysis of the evolutionary relationships of all the major groups of organisms alive today. The tree is a compilation of the research of many different scientists, headed by Dr. Joel Cracraft of the American Museum of Natural History. Each contributing scientist is an expert in one of the groups shown here. The data on which the analysis is based is a combination of anatomical, developmental, and genetic information. The tree is far more detailed than was possible to achieve in Darwin's day, yet it abundantly confirms his grand prediction that there is one—and only one—evolutionary "tree of life."

when many more, especially older fossils (older than 2.5 million years, that is) were found, it became apparent that the human evolutionary tree is more complex, more "bushy," than hitherto suspected—a fact that would have delighted the younger Charles Darwin especially since it shows how important isolation and speciation (and even turnovers) have been in both generating and conserving human evolutionary change. The recent announcement of tiny "Floresian man"—a vestige, apparently, of a species (*Homo ergaster*) that evolved on the African plains 1.8 million years ago, surviving to an astonishing 13,000 years ago in miniaturized form on the island of Flores in the Indonesian archipelago—is a stunning case in point. Floresian man highlights the importance of isolation in generating evolutionary change (miniaturization in this case), and the survival of truly ancient species.

But whether bush or tree, as a glance at the diagram on page 229, summarizing the main fossil evidence of human evolution over the past 5 million years, clearly shows, there was indeed an increase in brain size, as well as the advent of bipedalism and tool making (not shown on the diagram). A physicist at a conservative Christian college once said to me that I had "gone for the jugular" when I showed slide after slide of human fossils arranged in their correct order through time in a lecture the previous evening. The human fossil record is easily the creationists' worst nightmare.

And so, too, with the genetic evidence. How can anyone dismiss the obvious implication of the arresting fact that there is less than 2 percent difference between the genes of chimps and the genes of humans? Nor can those genes that are not the same really account for the differences between us and chimpanzees. The latest best thinking (as we have seen) is that this is more a matter of the timing of developmental events through the switching on and off of genes in fact held in common than it is purely a matter of the negligible genetic difference itself.

There's no getting around it: all the evidence puts us squarely within the ranks of apes, which are primates, which are mammals, which are animals, which are eukaryotes, which are a segment of *all life*. This is what we would expect if we evolved. This is what we see. We evolved. No doubt about it. Darwin was right.

Analysis of the human evolutionary fossil record by Dr. Ian Tattersall of the American Museum of Natural History. Very little was known of human fossils in Darwin's day. Subsequent intense exploration has revealed an extraordinarily rich and dense fossil record, showing speciation events, diversity of lineages, stasis—and a trend (as in horse evolution) to larger size and (especially significant to understanding the human condition) a larger brain through time.

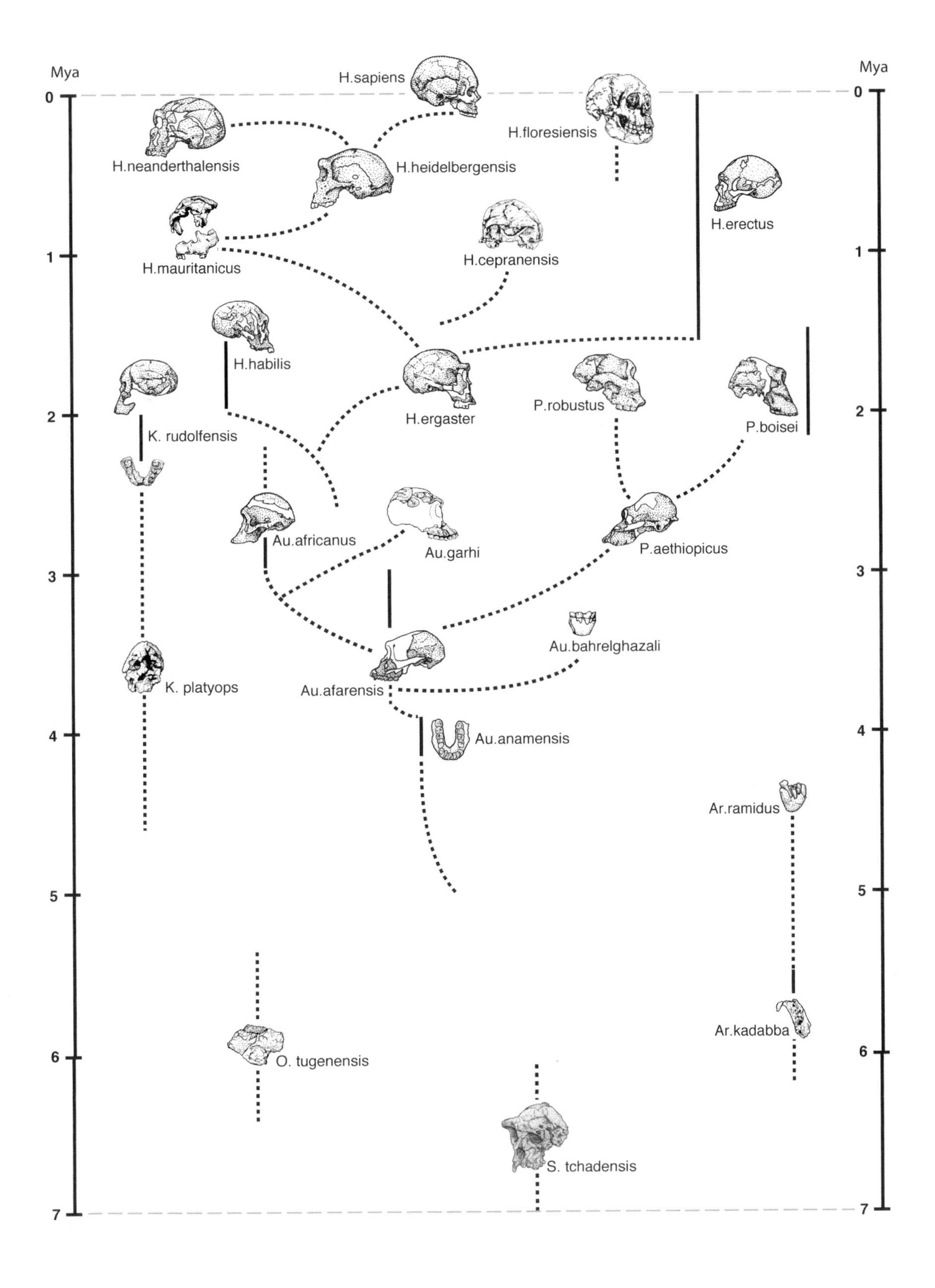

Mya
0
1
2
3
4
5
6
7
H.sapiens
H.floresiensis
H.neanderthalensis
H.heidelbergensis
H.erectus
H.mauritanicus
H.cepranensis
H.habilis
K. rudolfensis
H.ergaster
P.robustus
P.boisei
Au.africanus
Au.garhi
P.aethiopicus
Au.bahrelghazali
K. platyops
Au.afarensis
Au.anamensis
Ar.ramidus
Ar.kadabba
O. tugenensis
S. tchadensis

"Intelligent Design"

But despite all this evidence, all these predictions that confirm evolution over and over again, creationism is once again on the rise, coupled with the trend toward political and general religious conservatism that is apparent in many Judeo-Christian and Islamic societies. Even Italy, where secularism has long lived comfortably alongside the Catholic Church, now has its anti-evolutionists in high governmental, policy-making places. And Darwin's name is more often than not invoked instead of the more technical term "evolution": it's just easier to organize resistance to a person symbolizing the work of the devil than it is to characterize and refute a scientific theory.

Time was when creationists in the United States and elsewhere were more open and honest about their motives. The old "monkey laws" of Tennessee and some other states simply forbade the teaching of any doctrine that seemed at odds with the teachings of the Bible. But that movement eventually failed as the courts found clear violation of the First Amendment's provision for the separation of church and state. Even here, it was not until 1968 that the "monkey laws"—including Tennessee's—were actually found to be unconstitutional by the U.S. Supreme Court. The famous Scopes Trial of 1925 ended in overturning the conviction of biology teacher John Scopes on a technicality on appeal; the case never did reach the Supreme Court as Scopes's attorney Clarence Darrow had originally hoped.

In the 1960s, "Scientific Creationism" became the preferred strategy to sneak around First Amendment barriers to creationism in the classroom. The idea was to raise so many objections to evolution as a scientific theory that the legitimacy of creationism as a viable alternative would automatically be enhanced. The 1980 trial in Federal Court of Arkansas Statute 590, which was drafted to contrast the main points of evolutionary science and so-called creation science, was the high-water mark of this particularly deceitful and hypocritical venture. Before finding for the plaintiffs (all clergymen) who had challenged the law as an attack on quality science education in Arkansas, Judge William Overton brilliantly asked one of the witnesses testifying in support of the law why we would need a law to teach creation science if it were really science! It was good to see in this instance that blatant subterfuge was unable to pull the wool over the court's eyes.

Creationists of course go on attacking science, continuing to claim that the earth is really young (despite the misgivings that some have about "young-earth" creationism), and that all life appeared about the same time in fixed groups that may show variation inside the group but no evidence of connections between larger-scale groups like mammals and reptiles. But they have learned their lesson and generally have refrained from listing the tenets of creation science alongside those of modern geology and biology in legal documents.

Instead, in the latest version of their wolf-in-sheep's-clothing strategy to circumvent both the U.S. Constitution and normal canons of decency, creationists have assumed an even more innocent affect than the clumsy "Scientific Creationism" movement. My Internet service provider recently ran a news story entitled "School Mandates Alternate Evolution Theory." In a story that is all too distressingly familiar these days, the article recounts the successful efforts of some school board members in a small town in south-central Pennsylvania to mandate the teaching of this latest ploy: "Intelligent Design." The head of the curriculum committee—a born-again Christian and admitted creationist—nonetheless hypocritically maintains his purity of motive, simply to provide a "balanced presentation."

What is Intelligent Design? Nothing but William Paley's Watchmaker born again. The idea is that organisms are simply too complex to have been developed through natural processes; their very existence, like Paley's watch, implies they must be the handiwork of some Intelligent Designer. In a further shameless hypocritical indulgence, creationists will tell you with a straight face that, by making the case for an Intelligent Designer, they don't necessarily mean God. Creationists, faith in the gullibility of their fellow citizens seems boundless.

Complexity of biological structure is once again alleged to be evolution's fatal flaw, for no random process could assemble a human eye, let alone an entire human body. But evolution is not a random process: the mutations and the variations they produce might be "random"—but only with respect to what is a good or a bad thing for an organism to have. Natural selection is the powerful anti-chance culling device, the natural means by which harmful variants are discarded and only the useful kept. Evolution is indeed mindless; but it is far from random.

Then consider the development of a healthy human child from a single, microscopic fertilized egg. Darwin was proudest of his linking of development with evolution. The once mysterious process of development of a single fertilized cell into a body consisting of *billions* of cells of hundreds of different types has happened by the time every baby is born. Talk about complexity developing before our eyes!

And, thanks to the molecular revolution in biology, we now know much more about how that initial set of genetic instructions in the nucleus of the fertilized egg is translated as development proceeds: the mystery is beginning to recede in the light of this new, hard-won knowledge. We know experimentally that genes can be altered to change the outcome of the developmental process—offering hope for the correction of genetic diseases before infants are even born. Small changes in genetic information can profoundly alter the results of the developmental process: what a newborn will look like, how it will behave, what its genetically determined strengths and weaknesses will be.

Complexity is the outcome of a genetic program that can be altered. Evolution is simply the claim that, over billions of years of time, the genetic programs of organisms come to be altered through natural processes. We have observed those processes and now have the means to alter this information, and hence developmental pathways, ourselves.

None of this was known when Darwin was born in 1809; indeed, it wasn't even known in any detail by the time he died in 1882. Back then, Reverend naturalists like William Paley could make an impassioned plea that the human eye, like a watch, is so complex that there must be some equivalent of a Watchmaker—some Intelligence—assembling eyes, or indeed any and all organic structures. Darwin's contemporary and rival, Richard Owen, like most anatomists caught up in the complexities of the organisms they dissected, opposed evolution on these grounds. Even Darwin himself was persuaded by the argument—until he consciously grasped the evidence of his senses and began to think otherwise. As he remarked at one point later in life, the idea that God was continually correcting His own mistakes, removing some species and creating others later when He could have gotten the whole thing right in the first place, actually paints a most unflattering picture of the intelligence of the Intelligent Designer. But in light of what we know now about the actions and modifiability of genetic information, we no longer need such a quaint analogy to explain the complexity of organic structures.

There is a sort of dreamy Catch-22 quality about Intelligent Design. On the one hand, creationists are using the biological world as direct evidence of their claim that some Intelligent Designer is behind the whole thing. On the other hand, though their ideas purport to be scientific (or at least intellectually rigorous), they are conveniently untestable—since nearly everyone agrees that there is no way to experience the supernatural using the evidence of our senses as required by the basic rules of science. Even were the Intelligent Designer held *not* to be supernatural, but rather some sort of real Being exerting a real force in nature (other than natural selection), the very first line of investigation must surely be the completed demonstration of the existence of this Intelligent Designer.

Intelligent Design people don't want to go there. They say that biological complexity as we see it around us is evidence enough of the existence and actions of an Intelligent Designer. But sometimes, in another burst of false humility, they present their story as a "hypothesis"—albeit one that is as sufficiently supported by the evidence as is evolution. Equal time again. But despite some rather lame, inadequate, and little-publicized efforts to test the "Intelligent Design hypothesis," testability of their hypothesis is not at the top of the contemporary creationists' agenda. The textbook that the south-central Pennsylvania school board wants the students to read in biology is just the old plausible-alternative-that-you-ought-to-believe sort of thing.

Because Intelligent Design advocates have done such a poor job of formulating ways to test their hypothesis—and indeed because it would seem that the Intelligent Designer is in principle unknowable except through His works—I have recently developed a promising new avenue to test the very core of intelligent design. I use lower case for intelligent design here because I am speaking now not of ways of experiencing the supernatural but instead of asking the simple question: Are there any other intelligently designed systems that we *can* experience? And then the sixty-four thousand dollar question: How do such intelligently designed systems compare with biological systems?

Of course there are such systems. The computer on which I am typing these words is but one example of hundreds of thousands, probably millions, of types of designed artifacts—all the different sorts of objects, utilitarian and not, designed by hominids for at least the past 2.5 million years. To avoid the dicey (to creationists) subject of ancient hominids, and all that implies about human evolution, let's just focus on material cultural systems that are being made by humans today.

My premise is simple: If biological systems were created by an Intelligent Designer, what would we expect those systems to look like? If the history of designed systems looks a lot like the history of biological systems, then maybe, in good scientific fashion, we can say that, though we haven't proved the existence of an Intelligent Designer behind biological systems, we have failed to falsify the notion either. On the other hand, if biological and intelligently designed systems look very different from one another, we may, I think, claim to have cast serious doubt on the "hypothesis" of Intelligent Design insofar as biological systems are concerned.

I became interested in comparing biological and material cultural evolutionary patterns as a problem in its own right. I began my academic career as a fledgling anthropologist before shifting over to paleontology and evolutionary biology. There has been a tendency in scientific circles of late, in fact, to use the techniques of evolutionary biology to reconstruct human material cultural design history ("material cultural evolution").

In fact, the prevailing thinking has stressed the similarity between biological and cultural systems. Evolution might reasonably be defined as the fate of transmissible information in an economic context—a definition that fits both sorts of systems. And though people have realized all along that there are vast differences between the ways information is transmitted in the two systems—genes versus learning—nonetheless, simple models of selection working on variation have dominated the analyses so far. But just as a simple adaptation–through–natural selection vision yields an oversimplified biological evolutionary theory, reducing design evolution to a naive, "build a better mousetrap" selection in the marketplace oversimplifies the complexities of cultural evolution.

The complexity of material cultural systems in the modern world is easily on an order of magnitude greater than that of any known biological system. That in itself is no indication that an Intelligent Designer isn't behind the biological system; but it does mean that there is little detailed resemblance to be found when a biological system is compared with a human-designed system. The differences arise for the most part from two aspects of the origin and transmission of information in designed systems when compared with genetic systems.

To explore these differences between material cultural and biological systems, I have developed a database on the history of piston-valved cornets: brass musical instruments invented in 1825 and still made today. The full database consists of 17 variables describing 123 distinct cornet models made by nearly 200 different makers. The form of the database is exactly like those I and my colleagues routinely produce for our organisms (in my case, trilobites, an extinct group of the arthropod phylum).

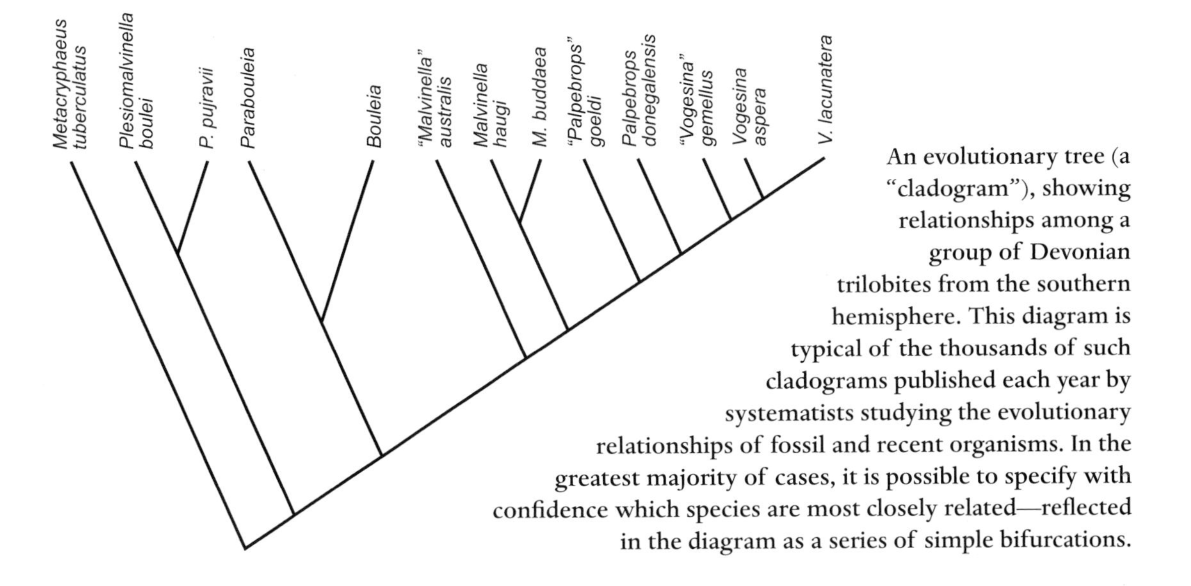

An evolutionary tree (a "cladogram"), showing relationships among a group of Devonian trilobites from the southern hemisphere. This diagram is typical of the thousands of such cladograms published each year by systematists studying the evolutionary relationships of fossil and recent organisms. In the greatest majority of cases, it is possible to specify with confidence which species are most closely related—reflected in the diagram as a series of simple bifurcations.

Compare the two diagrams on pages 235–36. The first is an evolutionary tree ("cladogram") produced from trilobite data. It shows that in the vast majority of cases, the computer can locate the nearest match for any species, resulting in a neat array of dichotomous branches that show the pattern of evolutionary relationships—which species is most closely related to which—and how that group of two fits in with the other species represented in the database.

Now look at the diagram generated from the cornet database—a diagram for simplicity's sake that is based on thirty-nine basic cornet models. The degree of resolution is much lower than in the trilobite diagram; there is no easy telling what model of cornet is more closely "related" to what. This is because there is almost no limit to the degree of "mix-and-match," the permutations and combinations of the seventeen different cornet features. Information can be transmitted "horizontally" between models—meaning that inventions that come along later can be applied retroactively to older models. Nothing like that is found in biological systems, except in bacteria.

Adding to the confusion in design history is what I have called the "Hannah Principle" (derived from the thinking of the contemporary industrial designer Bruce Hannah of Pratt Institute). In biological evolution, new structures arise from old: the limbs of four-legged land-living vertebrates are modified versions of fish fins (lobe-finned fishes,

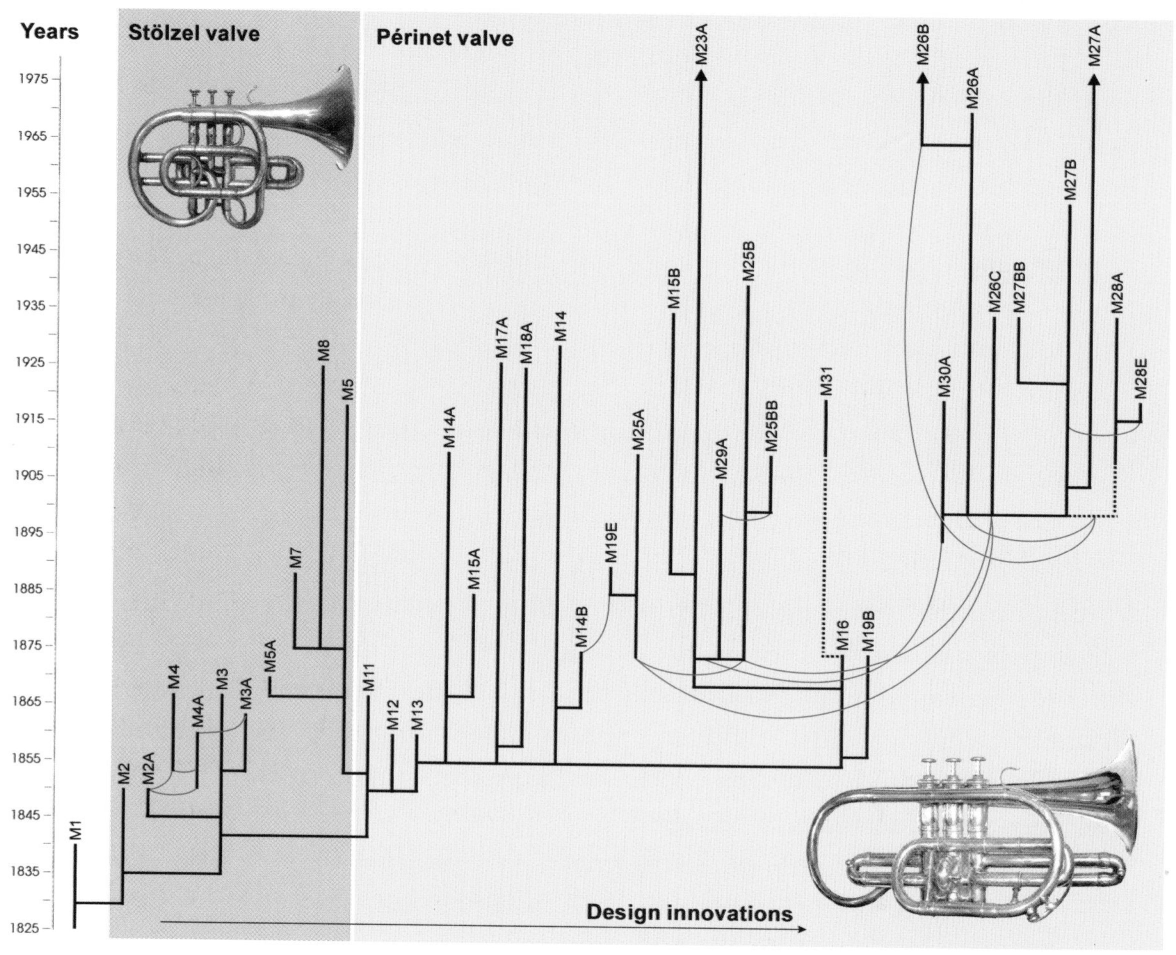

A cladogram of cornet evolutionary history, derived from the same basic techniques used to analyze biological evolutionary history. Note the lack of resolution into simple bifurcations: intentionally designed systems, unlike biological systems, show the spread of information between "lineages," creating a complex diagram of historical relationships that looks quite different from a typical cladogram of biological evolutionary history. The differences between the diagrams of biological and human-designed systems challenge the notion of "Intelligent Design"—the claim that biological systems by their very nature imply the existence of a conscious Designer.

not the spiny thin fins of most modern fish), and the wings of birds, bats, and flying reptiles (pterosaurs) are further modifications of the forelimb of land-living reptiles and mammals (in the case of bats).

Not so in many examples of material cultural design evolution, where new versions of an old object as often represent independently invented solutions to common problems (the Hannah Principle). The "evolution" of the Périnet valve (the type of piston valve used on cornets and trumpets today) really represents a succession of alternate designs produced by different makers in a competitive market, in some instances to get around the limitations imposed by patents. In this respect, design evolution is wholly unlike biological evolution.

Though an Intelligent Designer is represented as a single mind and not a mélange of different competing factories busy stealing ideas from each other, nonetheless, this simple demonstration of the vast difference between both the processes and the results of material cultural and biological evolution does, I think, cast serious doubt on the notion of Intelligent Design lurking behind biological systems.

The funny thing is that there is actually a creationist who has said as much: Gary Parker, in his book *Creation: The Facts of Life* (1980). Parker takes the astonishing position that there really is no nested set of resemblances linking up all forms of life. He paints instead a picture like the one I have just presented for my cornets: life as a mishmash of features, with no order to it, a sort of permutations-and-combinations array of features that one might expect from the fertile mind of the Creator. Of course God would have wanted to apply a good idea invented for one set of organisms to other sets as well. Parker denies transformation of characters, invoking instead the lateral transfer of information and the Hannah Principle in the mind of God as He created life. After all, Parker says, that's what you would expect from an all-powerful Creator.

Exactly so. And that is precisely what we do see in the details of the "evolution" of a designed system such as cornets. But it is emphatically *not* what we see in the history of life.

So much for taking creationism sufficiently seriously to formalize comparisons between biological and material cultural systems to test the "hypothesis" of Intelligent Design. It is sad that it still comes down to this—sad that so many people still feel so threatened by the idea that they are connected to the universe in the most fundamental possible sense.

One of the more arresting things I have read about Charles Darwin's life actually

Darwin in old age. Few people before or since have proved so original, as well as so industrious and productive for so many years, as Charles Robert Darwin had been throughout his life.

comes from his wife, Emma. Recall that Darwin, despite his father's advice against it, decided to come clean to his fiancée, cousin Emma, regarding the growing religious doubts stirred up in his mind by his work in the late 1830s. It was not Charles's doubts per se that troubled Emma so much as the disturbing thought in *her* mind that his doubts might mean that they would not spend eternity together. Medicine was primitive and life was short in those days—perhaps less so for the landed gentry, on average, than for the poor, but in any case short.

One definition of consciousness is the awareness of mortality, and so far as anyone knows we are the only species whose individuals are conscious of their own eventual death. This is quite a price to pay for the fantastic capacity to be aware of one's existence and to have the privilege of trying to make sense of the world—and of life—while we are here. Charles Darwin, like everyone else, was very fearful of his health and the thought of his eventual death, though, like so many others, he was in fact calm and accepting when his health did fail and death overtook him. But, scared as he was of his own mortality, and petrified to the point of immobility for twenty years as he harbored his dark, secret evolutionary thoughts ("like confessing a murder"), Darwin conquered his fears and eventually told the world what he thought, and why. The revolution he precipitated was complete in science soon after the *Origin* appeared in 1859. That it is still only half-complete in society at large is perhaps no great surprise.

Would that all of us were as courageous as Charles Robert Darwin.

BIBLIOGRAPHY

Charles Darwin: Books, Articles, Subsequently Published Manuscripts, and Correspondence

Anon. *The Darwin Correspondence Online Database.* http://libpro13.1ib.cam.ac.uk

Barlow, N., ed. *Darwin's Ornithological Notes. Bulletin of the British Museum (Natural History) Historical Series*, Vol. 2, no. 7, 1963.

Barrett, Paul H., P. J. Gautrey, S. Herbert, D. Kohn, and S. Smith, eds. *Charles Darwin's Notebooks. 1836–1844.* Ithaca: Cornell University Press, 1987.

Darwin, C., and A. R. Wallace. *Evolution by Natural Selection*, Foreword by Sir Gavin De Beer. Charles Darwin's 1842 *Sketch,* 1844 *Essay* (Introduction by Francis Darwin), and the three-part paper by Darwin and Alfred Wallace, *On the Tendency of Species to Form Varieties; and on the perpetuation of Varieties and Species by Natural Means of Selection.* Cambridge: University Press, 1958.

Darwin, C. *Journal of Researches into the Geology and Natural History of the Various Countries Visited by H.M.S.* Beagle, *under the Command of Captain FitzRoy, R.N. from 1832 to 1836.* London: Henry Colburn, 1839; 2nd ed., 1845.

———, ed. *Zoology of the Voyage of H.M.S.* Beagle *under the Command of Captain FitzRoy,* edited and superintended by Charles Darwin. Part 1. *Fossil Mammalia*, by Richard Owen. With a Geological Introduction by Charles Darwin. Part 2. *Mammalia*, by George R. Waterhouse. With a notice of their habits and ranges by Charles Dar-

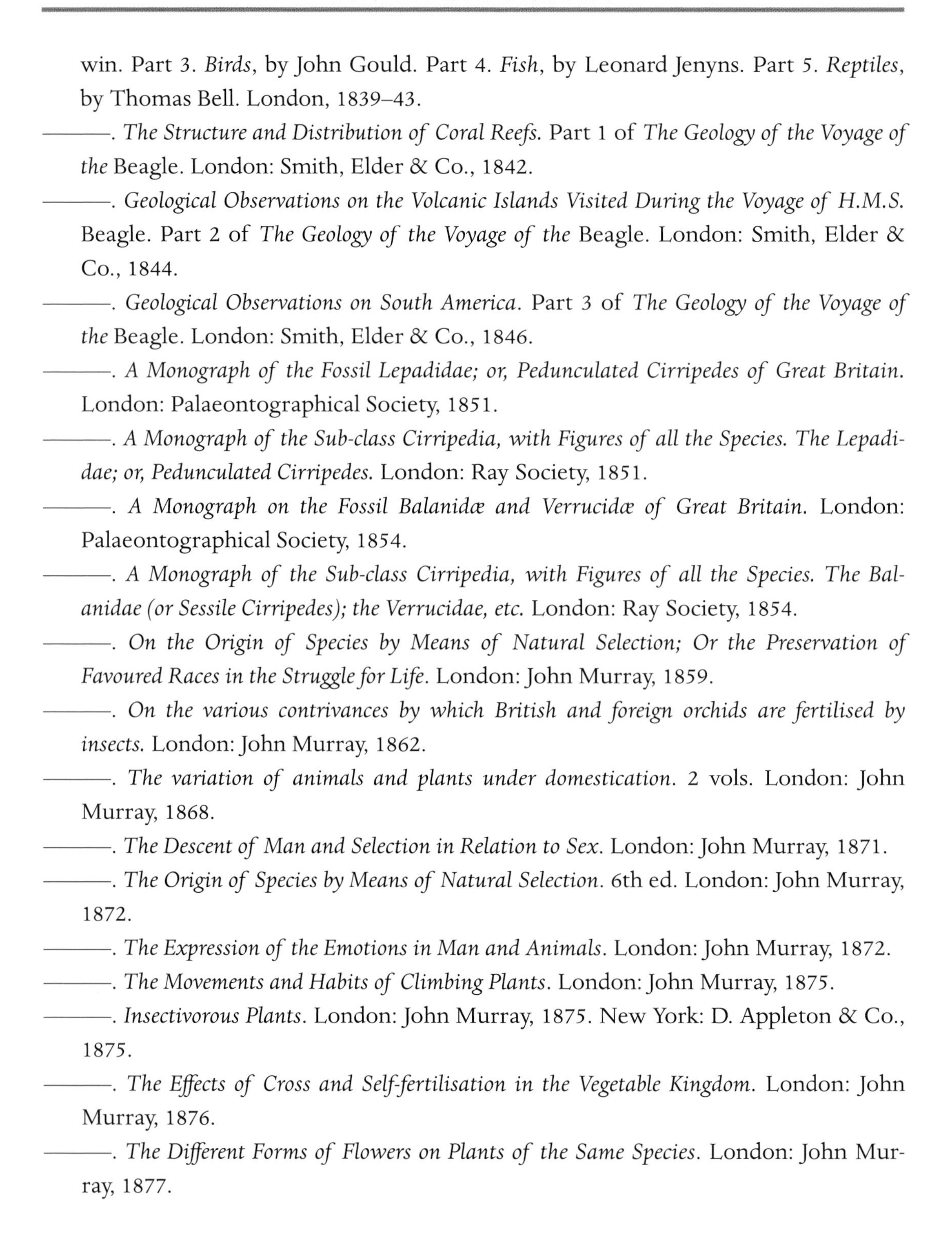

win. Part 3. *Birds*, by John Gould. Part 4. *Fish*, by Leonard Jenyns. Part 5. *Reptiles*, by Thomas Bell. London, 1839–43.

———. *The Structure and Distribution of Coral Reefs*. Part 1 of *The Geology of the Voyage of the* Beagle. London: Smith, Elder & Co., 1842.

———. *Geological Observations on the Volcanic Islands Visited During the Voyage of H.M.S.* Beagle. Part 2 of *The Geology of the Voyage of the* Beagle. London: Smith, Elder & Co., 1844.

———. *Geological Observations on South America*. Part 3 of *The Geology of the Voyage of the* Beagle. London: Smith, Elder & Co., 1846.

———. *A Monograph of the Fossil Lepadidae; or, Pedunculated Cirripedes of Great Britain*. London: Palaeontographical Society, 1851.

———. *A Monograph of the Sub-class Cirripedia, with Figures of all the Species. The Lepadidae; or, Pedunculated Cirripedes*. London: Ray Society, 1851.

———. *A Monograph on the Fossil Balanidæ and Verrucidæ of Great Britain*. London: Palaeontographical Society, 1854.

———. *A Monograph of the Sub-class Cirripedia, with Figures of all the Species. The Balanidae (or Sessile Cirripedes); the Verrucidae, etc.* London: Ray Society, 1854.

———. *On the Origin of Species by Means of Natural Selection; Or the Preservation of Favoured Races in the Struggle for Life*. London: John Murray, 1859.

———. *On the various contrivances by which British and foreign orchids are fertilised by insects*. London: John Murray, 1862.

———. *The variation of animals and plants under domestication*. 2 vols. London: John Murray, 1868.

———. *The Descent of Man and Selection in Relation to Sex*. London: John Murray, 1871.

———. *The Origin of Species by Means of Natural Selection*. 6th ed. London: John Murray, 1872.

———. *The Expression of the Emotions in Man and Animals*. London: John Murray, 1872.

———. *The Movements and Habits of Climbing Plants*. London: John Murray, 1875.

———. *Insectivorous Plants*. London: John Murray, 1875. New York: D. Appleton & Co., 1875.

———. *The Effects of Cross and Self-fertilisation in the Vegetable Kingdom*. London: John Murray, 1876.

———. *The Different Forms of Flowers on Plants of the Same Species*. London: John Murray, 1877.

———, assisted by Francis Darwin. *The Power of Movement in Plants*. London: John Murray, 1880.

———. *The Formation of Vegetable Mould, Through the Action of Worms, with Observations on Their Habits*. London: John Murray, 1881.

Darwin, F., ed. *The Foundations of the Origin of Species. Two Essays in 1842 and 1844 by Charles Darwin*. Cambridge: University Press, 1909.

———. *Charles Darwin's Autobiography. With his notes and letters depicting the Growth of the Origin of Species*. New York: H. Schuman, 1950.

De Beer, G., ed. *Darwin's Notebooks on Transmutation of Species*. Parts I–IV + *Addenda and Corrigenda. Bulletin of the British Museum (Natural History), Historical Series*, Vol. 2, nos. 2–6, 1960–61.

Keynes, R., ed. *The* Beagle *Record. Selections from the Pictorial Records and Written Accounts of the Voyage of H.M.S.* Beagle. Cambridge: University Press, 1979.

———. *Charles Darwin's Zoology Notes and Specimen Lists from H.M.S.* Beagle. Cambridge: University Press, 2000.

Stauffer, R. C., ed. *Charles Darwin's Natural Selection. Being the Second Part of His Big Species Book Written from 1856 to 1858*. Cambridge: University Press, 1975.

van Whye, J. *The Writings of Charles Darwin on the Web* (n.d.) http://pages.britishlibrary.net/charles.darwin/

Books and Articles on Charles Darwin

Browne, Janet. *Charles Darwin. Voyaging. A Biography*. New York: Alfred A. Knopf, 1995.

———. *Charles Darwin. The Power of Place*. Vol. II of *A Biography*. New York: Alfred A. Knopf, 2002.

Desmond, A., and J. Moore. *Darwin. The Life of a Tormented Evolutionist*. New York: W. W. Norton, 1991.

Ghiselin, Michael T. *The Triumph of the Darwinian Method*. Berkeley and Los Angeles: University of California Press, 1969.

Gruber, Howard E. *Darwin on Man. A Psychological Study of Scientific Creativity*. Chicago: University of Chicago Press, 1974, 1981.

Hull, D. L. *Darwin and His Critics*. Chicago and London: University of Chicago Press, 1973.

Keynes, Randal. *Annie's Box. Charles Darwin, His Daughter, and Human Evolution*. London: Fourth Estate, 2001. Published in the United States as *Darwin, His Daughter and Human Evolution*. New York: Riverhead Trade, 2002.

Keynes, Richard. *Fossils, Finches and Fuegians. Charles Darwin's Adventures and Discoveries on the* Beagle, *1832–1836*. London: HarperCollins, 2002.

Kohn, D. "Theories to Work By: Rejected Theories, Reproduction and Darwin's Path to Natural Selection," *Studies in the History of Biology*, 4 (1980), 67–170.

Sulloway, F. J. "Geographic Isolation in Darwin's Thinking: The Vicissitudes of a Crucial Idea," *Studies in the History of Biology*, 3 (1979), 23–65.

Other Scientific and Historical Books and Papers

Brosius, J. "Disparity, Causation, Adaptation, Exaptation, and Contingency at the Genome Level," *Paleobiology*, 31 (2, Supplement, 2005), 1–16.

Chambers, R. *Vestiges of the Natural History of Creation*. Edinburgh, 1844.

Cuvier, G. *Discours sur les révolutions de la surface du globe*. Paris, 1812.

Dobzhansky, Theodosius. *Genetics and the Origin of Species*. New York: Columbia University Press, 1937.

Eldredge, N. "The Allopatric Model and Phylogeny in Paleozoic Invertebrates," *Evolution*, 25 (1971), 156–67.

———, and S. J. Gould. "Punctuated Equilibria: An Alternative to Phyletic Gradualism," in T. J. M. Schopf, ed., *Models in Paleobiology*. San Francisco: Freeman, Cooper, 1972.

Eldredge, N. *The Pattern of Evolution*. New York: W. H. Freeman, 1999.

———. *The Triumph of Evolution and the Failure of Creationism*. New York: W. H. Freeman, 2000.

———, et al. "The Dynamics of Evolutionary Stasis," *Paleobiology*, 31 (2, Supplement, 2005), 133–45.

Gould, S. J. "Is Uniformitarianism Necessary?" *American Journal of Science*, 263 (1965), 223–28.

Grant, P. R., and B. R. Grant. "Adaptive Radiation of Darwin's Finches," *American Scientist*, 90 (2002), 131–39.

———. "What Darwin's Finches Can Teach Us About the Evolutionary Origin and Regulation of Biodiversity," *Bioscience*, 53 (2003), 965–75.

Gregory, T. R. "Macroevolution, Hierarchy Theory, and the C-Value Enigma," *Paleobiology*, 30 (2004), 179–202.

Herschel, J. F. W. *A Preliminary Discourse on the Study of Natural Philosophy*. London, 1830.

Lyell, Charles. *Principles of Geology*. 3 vols. London: John Murray, 1830–33.

———. *The Geological Evidences of the Antiquity of Man, with Remarks on the Theories of the Origin of Species by Variation*. London: John Murray, 1863.

Malthus, T. R. *An Essay on the Principle of Population, as it Affects the Future Improvement of Society*. London: J. Johnson, 1798. [Darwin read the sixth edition, 1826]

Maresca, B., and J. Schwartz. "Environmental Change and Stress Protein Concentration as a Source of Morphological Novelty: Sudden Origins. A General Theory on a mechanism of Evolution," *Evolution and Development*, submitted (n.d.).

Mayr, Ernst. *Systematics and the Origin of Species*. New York: Columbia University Press, 1942.

Mendel, J. (Gregor). "Versuche über Pflanzen-hybriden," *Verhandlungen des Naturforschenden Vereins Brünn*, 4 (1866), 3–57.

Paley, W. *Natural Theology: Or, Evidences of the Existence and Attributes of the Deity, Collected from the Appearances of Nature*. London: R. Fauldner, 1802.

Rowe, A. W. "An Analysis of the Genus *Micraster*, as Determined by Rigid Zonal Collecting from the zone of *Rhynchonella Cuvieri* to that of *Micraster cor-anguinum*," *Quarterly Journal of the Geological Society of London*, 55 (1899), 494–547.

Simpson, G. G. *Tempo and Mode in Evolution*. New York: Columbia University Press, 1944.

Tattersall, I. *Becoming Human: Evolution and Human Uniqueness*. New York: Harcourt, 1998.

Vrba, E. S. "Mass Turnovers and Heterochrony Events in Response to Physical Change," *Paleobiology*, 31 (2, Supplement, 2005), 157–74.

Whewell, William. *History of the Inductive Sciences*. London: Parker, 1837.

INDEX

Page numbers in *italics* refer to illustrations.